Nelson N Titus

The American Eclectic Practice of Medicine, as Applied to the Diseases of Domestic Animals

In which every form of disease peculiar to animals is treated in accordance with the most approved methods of modern science

Nelson N Titus

The American Eclectic Practice of Medicine, as Applied to the Diseases of Domestic Animals

In which every form of disease peculiar to animals is treated in accordance with the most approved methods of modern science

ISBN/EAN: 9783337240523

Printed in Europe, USA, Canada, Australia, Japan

Cover: Foto ©berggeist007 / pixelio.de

More available books at **www.hansebooks.com**

THE AMERICAN

ECLECTIC PRACTICE OF MEDICINE,

AS APPLIED TO THE

DISEASES OF DOMESTIC ANIMALS;

IN WHICH EVERY FORM OF DISEASE PECULIAR TO ANIMALS IS TREATED IN ACCORDANCE WITH THE MOST APPROVED METHODS OF MODERN SCIENCE.

WITH A TREATISE ON

BLEEDING, BLISTERING, MERCURIALIZING,

AND THE

PERNICIOUS EFFECTS OF MINERALS ON THE ANIMAL FIBERS.

WITH

A LIST OF POISONS AND THEIR ANTIDOTES.

TOGETHER WITH A

HISTORICAL SKETCH OF THE HORSE,

AND THE

PRINCIPLES OF BREEDING, REARING, TRAINING, AND THE GENERAL MANAGEMENT OF THE HORSE.

BY

NELSON N. TITUS.

———◆———

NEW YORK:
BAKER & GODWIN, PRINTERS,
Printing-House Square, opposite the City Hall.
1862.

Entered according to Act of Congress, in the year 1861, by

NELSON N. TITUS,

In the Clerk's Office of the District Court of the United States for the Northern District of New York.

PREFACE.

THE object of this publication is to render as plain and familiar as possible a subject that has been involved in much obscurity. The want of a work possessing practical facts and illustrations, on the Eclectic System, for the cure of disease in domestic animals, has long been severely felt and acknowledged. Under this conviction, and at the solicitation of my friends, I was induced to lend my aid in bringing forth the present volume.

This work contains, in a concentrated form, all the useful information with regard to the treatment of disease in domestic animals that could be gained from any source. I have labored incessantly to make myself familiar with all the medical systems extant, and more especially the Botanic and Eclectic systems. From this store of knowledge, which I have accumulated by study, observation, and an extensive practice of ten years in the diseases of domestic animals, and twelve years as an eclectic physician, I have been enabled to bring before the American public a work which I sincerely believe will prove extremely serviceable in the treatment of the diseases of domestic animals. It is divested of all technicalities, and written in plain language, that any one can understand. It is a practical, domestic work, and calculated principally for the use of the farmer. I am not aware of any publication in which the veterinarian science, or the art of farriering, has been laid down in such a manner as to be clearly understood. This work is so simple and familiar in its composition as to render it at once intelligible to every one who may think proper to peruse its pages, and is the first eclectic work published on this subject.

All works that I have seen on the diseases of domestic animals, are not only worthless but injurious in their effects; they all appear to have been drawn from medical works on the human subject; they have been written by professed scientific men, who in reality know little or nothing of the laws that govern the animal economy or the physiology of the horse and other animals. For this reason they have supposed that they were subject to nearly all the diseases incident to mankind. Hence they have multiplied, classified, divided, and subdivided diseases almost without number—diseases that exist nowhere but in the imagination of the authors; hence their *materia medica* has been filled with a mass of acrid, drastic, mineral compounds, which always produce effects worse than the original disease.

It is a well-known fact, drawn from observations and experience, that what is harmless to one species of animals is poison to others. For instance, the laurel is poison to sheep, and harmless to the deer and round-horned elk; pepper is poison to the swine, which to man is a useful spice; arsenic is poison to mankind and many species of animals, yet it can be given to the horse in doses of one or two ounces, without producing any sensible or immediate effects; and many articles which are harmless to man are poison to the horse, many of which articles they have in their pharmacy. For this reason you will readily perceive that when they draw their conclusions and prescriptions from medical works they commit a great blunder. I desire particularly to observe that all of my selections of medicines are from the vegetable kingdom.

In this work I have paid little regard to the opinions or theories of others, but have based my system mainly upon practical facts, observations, and experience. In all works written on the subject of medicine, one author borrows from another, not only ideas, but prescriptions; so you get little or nothing original, consequently, no improvement.

This volume contains a treatise on Bleeding, Blistering, and Mercurializing, with the pernicious effects of minerals in general, and a list of Poisons and their Antidotes.

Also the principles of Breeding, Rearing, Training, and the general Management of the Horse; with a historical sketch of the Horse, and more especially the American Trotting Horse, the peculiarities of this breed, and a record of their time; their height; rules to be observed in purchasing a horse; the secret of Horse-Taming; together with the secret of training the Horse for the course; and, in addition to all, my most valuable receipts on the diseases of animals—receipts that I have been twenty years in collecting, for which neither pains nor money have been spared—receipts worth more than one hundred times the price of this work, and which I have repeatedly been solicited to sell, but have reserved for the benefit of those that purchase this work.

HISTORY OF THE HORSE.

The Horse, it seems, is an inhabitant of the eastern continents only; no trace of it having been met with as showing that it existed in any part of America until it was carried thither by the European settlers. In the wild plains of South America, immense herds of wild horses are met with, which are descended from the Andalusian breed, originally conveyed from Spain by the first conquerors; and these are the most frequently found in the southern districts of the river Plata, as far as Rio Negro, the country of the Patagonians, and the districts immediately adjoining. Some of the herds amount to no less than ten thousand animals, each troop comprising many families.

The great tracts of desert country around the Sea of Aral and the Caspian Sea, have been supposed to be the native residence of the Horse; but if this conjecture be correct, the animal must have widely extended his geographical range, for he is found in a wild state in Asia, as far north as the sixteenth degree, and to the utmost southern extremes of that vast continent, and also in many parts of Africa. So late as the seventh century of the Christian era, when the prophet Mahomet attacked the Kerish, not far from Mecca, he had but two horses in his train; and although in the plunder of his horrible campaign he carried with him in his retreat twenty-four thousand camels, forty thousand sheep, and twenty-four thousand ounces of silver, there is no mention of horses being part of the booty. We are informed that the Arabians then had but few horses, and those not at all valued; so that Arabia, whence are now the most celebrated coursers in the world, is comparatively of but modern date as a breeding country.

Geological researches, however, have discovered fossil remains of the Horse in almost every part of the world, "from the tropical plains of India to the frozen regions of Siberia—from the northern extremities of the New World to the southern points of America." But among the Hebrews horses were rare previous to the days of Solomon, who had horses brought out of Egypt after his marriage with the daughter of Pharaoh; and so rapidly did he multiply them, by purchase and by breeding, that those kept for his own use required, as it is written, "forty thousand stables, and forty thousand stalls." Hence, when honored by a visit from the beautiful Queen of Sheba, bringing with her "camels bearing spices," and "very much gold and precious stones," it was doubtless in the contemplation of his magnificent stud of horses and chariots, kept for the amusement of wives and concubines, as well as of his other vast displays of power and magnificence, that Her Majesty exclaimed, in the fullness of her admiration, "Behold! the half was not told me!"

This gallant monarch appears to have enjoyed a large monopoly of horse trade with Egypt, for which he was probably indebted to his having an Egyptian princess for one of his wives. His merchants supplied horses in great numbers to the Hittite kings of northern Phœnicia.

The fixed price was one hundred and fifty shekels for a horse, and six hundred shekels for a set of chariot horses.

We read, when Joseph proceeded with his father's body from Egypt into Canaan, "there accompanied him both chariots and horsemen" (Gen. l. 9); and the Canaanites are said to have gone out to fight against Israel "with many horses and chariots" (Joshua xi. 4). This was more than sixteen hundred years before Christ.

The Horse was employed on the course as early as 1450 B. C., when the Olympic games were established in Greece, at which horses were used in chariot and other races.

Professor Low says: "The Horse is seen to be affected in his character and form by the agency of food and climate, and it may be, other causes unknown to us. He sustains the temperature of the most burning regions; but there is a degree of cold at which he cannot exist, and as he approaches this limit his temperament and external conformation are affected. In Iceland, at the Artic Circle, he has become a dwarf; in Lapland, at latitude 65°, he has given place to the reindeer; and in Kamtschatka, at 62°, he has given place to the dog. The nature and abundance of his food, too, greatly affect his character and form. A country of heaths and innutricious herbs will not produce a horse so large and strong as one of plentiful herbage; the horse of mountains will be smaller than that of the watered valley."

There are many varieties of breeds of the Horse. Ineffectual attempts have been made to decide which variety now existing constitutes the original breed,—some contending for the Barb, and others for the wild horse of Tartary.

The principal breeds of horses now bred in the United States are the Race Horse, the Arabian, the Morgan, the Canadian, the Norman, the Cleaveland Bay, the Conestoga, the Virginia Horse, the Clydesdale, and the Wild or Prairie Horse.

THE RACE HORSE.

About this breed there is much dispute. Mr. Youatt says: "With regard to the origin of the *Thorough-bred Horse*, by some he is traced through both sire and dam to Eastern parentage; others believe him to be the native horse, improved and perfected by judicious crossings with the Barb, the Turk, or the Arabian. The Stud Book, which is an authority with every English breeder, traces all the old racers to some Eastern origin, or it traces them until the pedigree is lost in the uncertainty of an early period of breeding. Whatever may be the truth as to the origin of the Race Horse, the strictest attention has, for the last fifteen years been paid to pedigree. In the descent of almost every modern racer not the slightest flaw can be discovered."

The Racer is generally distinguished, according to the same authority, by his beautiful Arabian head; his fine and finely-set neck; his oblique shoulders; his well-bent hinder legs; his fine, muscular quarters; his flat legs, rather short from the knee downward, although not always so deep as they should be, and his long and elastic pastern.

The thorough-bred and half-bred horses are much better for domestic use, beside being much handsomer than the common horse, and their speed and power of endurance are much greater. The superiority of the Northern carriage and draught stock is owing to the fact that thorough-bred horses have found their way North and East from Long Island and New Jersey, where great numbers are annually disposed of that are unsuited to the course.

For the farm, the pure Thorough-bred Horse would be nearly useless. He lacks weight and power for the draught. For road-work, we have the same objection, although not to the same extent, perhaps. The best English road-horse is a cross of the Thorough-bred and the Cleaveland.

THE ARABIAN HORSE.

This is one of the most pure breeds of horses known in the world, and great care is taken that they are not mixed or contaminated by impure blood or with an inferior breed. The genealogy of the Arabian Horse, according to Arabian accounts, is known for two thousand years. Many of them have written and attested pedigrees, extending more than four hundred years; and, with true Eastern exaggeration, traced by oral tradition from the stud of Solomon. A more careful account is kept of these genealogies than of those of the most ancient family of the proudest Arab chief, and very singular precautions are taken to prevent the possibility of fraud, so far as the written pedigree extends.

Richardson says: " Often may the traveler in the desert, on entering within the folds of a tent, behold the interesting spectacle of a magnificent courser extended upon the ground, and some half-dozen little, dark-skinned, naked urchins scrambling across her body, or reclining in sleep, some upon her neck, some on her body, and others pillowed upon her heels; nor do the children ever experience injury from their gentle playmate. She recognizes the family of her friend, her patron, and toward them all the natural sweetness of her disposition leans, even to overflowing."

THE MORGAN HORSE.

Mr. S. W. Jewett, a celebrated stock-breeder, in an article in the Cultivator, says:—

"I believe the Morgan blood to be the best ever infused into the Northern horse, and is probably a cross between the English race-horse and the common New-England mare. It is,

THE ARABIAN HORSE.

perhaps, the finest breed for general usefulness now existing in the United States. The Morgans are well known, and esteemed for activity, hardiness, gentleness, and docility; and are well adapted for all kinds of work; good in all places, except for races on the turf. They are lively and spirited, lofty and elegant in their action, carrying themselves gracefully in the harness. They have clean bones, sinewed legs, compactness, short, strong backs, powerful lungs, strength and endurance. They are known by their short, clean heads, wide across the face at the eyes, eyes lively and prominent; they have width under the jaws, large windpipe, deep brisket, heavy and round body, broad quarters, a lively, quick action, with dark, flowing, wavy mane and tail. They make the best of roadsters, and live to a great age."

The celebrated *Sherman Morgan*, according to Linsley, was foaled in 1835, the property of Moses Cook, of Campton, N. H., sired by Sherman, g. sire Justin Morgan. The pedigree of the dam not fully established, but conceded to have been a very fine animal, and said to be from Justin Morgan. Sherman Morgan is fifteen hands high, weighs about 1,050 lbs., is dark chestnut, and very much resembles his sire, Sherman, but heavier, stockier, and not as much action. He is a fine horse, and is now kept at Lancaster, N. H., where Sherman died. He is owned by A. J. Congdon.

THE CANADIAN HORSE.

The Canadian Horse is mainly of Norman French descent. It is a hardy, long-lived animal, is easily kept, and very useful on a farm, although too small for heavy work. A cross between stallions of this breed and our common mares produces a superior horse, and such cross is finding favor among farmers.

THE NORMAN HORSE.

Mr. Harris, admiring the speed, toughness, and endurance of the French coach-horses, has resolved to import this valuable stock, and deserves the thanks of the American public for his perseverance and sacrifice in this enterprise. The Norman horses are enduring and energetic beyond description, and keep their condition on hard fare and brutal treatment, when most other breeds would suffer and die. This variety of horse is em-

ployed in France to draw the ponderous stage-coaches called
"diligences," and travelers express astonishment at the extraordinary performances of these animals. Each of these huge vehicles is designed for eighteen passengers, and, when thus
loaded, are equal to five tons weight. Five horses are attached
to the clumsy and cumbrous carriage, with rude harness, and
their regular rate of speed, with this enormous load, is seven
miles an hour, and this pace is maintained over rough and hilly
regions. On some routes the road is lighter, when the speed is
increased to eight, and sometimes ten, miles an hour.

The Norman Horse is from the Spanish, of Arabian ancestors, and crossed upon the draught-horse of Normandy.

CLEAVELAND BAY.

The Cleaveland Bay, according to Mr. Youatt, is nearly extinct in England. They were formerly employed as a heavy,
slow coach-horse. Mr. Youatt says: "The origin of the better kind of coach-horse is the Cleaveland Bay, confined principally to Yorkshire and Durham, with perhaps Lincolnshire on
one side and Northumberland on the other, but difficult to meet
with pure in either county.

The Cleaveland mare being crossed by a three-fourths of
thorough-bred, of sufficient height, but not of so much substance, we obtain the four-in-hand and superior curricle-horse.

Cleaveland Bays were imported into western New York a
few years since, where they have spread considerably. They
have often been exhibited at our State fairs. They are monstrously large, and, for their size, are symmetrical horses, and
possess considerable action; but they do not endure on the
road at any moderate pace. Whether they spring from the
genuine and unmixed Cleaveland stock, we have no means of
knowing. The half-bloods, the produce of a cross with our
common mares, are liked by many of our farmers. They make
strong, serviceable farm-horses, though rather prone to sullenness of temper.

THE CONESTOGA HORSE.

This Horse is more remarkable for endurance than symmetry, and is found chiefly in Pennsylvania and the adjacent
States. In height it seldom reaches seventeen hands, but the

legs are long and the body light, and they make very good carriage and heavy draft horses.

THE CLYDESDALE HORSE.

This breed of horses has descended from a cross between the Flemish horse and the Lanarkshire (Scotland) mares.

The name is derived from the district on the Clyde where the breed is chiefly found. Horses of this breed are well adapted for the cart and plow on heavy soil. They are strong, hardy, steady, true pullers, of sound constitutions, and from fourteen to sixteen hands high. They are broad, thick, heavy, compact, and well made for durability, health, and power. They have sturdy legs, strong shoulders, back, and hips, a well-arched neck, and a light face and head.

THE VIRGINIA HORSE.

The Virginia Horse abounds to a greater or less extent in all the Southern, Western, and Middle States. It derived its origin from English blood-horses imported at various times, and has been most diligently and purely kept in the South. The celebrated horse Shark, the best horse of his day, was the sire of the best Virginia horse, while Tally-ho, son of Highflyer, peopled the Jerseys.

THE WILD PRAIRIE HORSE.

The Wild Prairie Horse has doubtless sprung from the Spanish stock, like the wild horse of the pampas and other parts of the Southern Continent, all of which are of the celebrated Andalusian breed, derived from the Moorish Barb. The Prairie Horse is often captured, and when domesticated is found to be capable of great endurance. He is not, however, very elegant, nor very symmetrical, being rather small and scrubby.

THE AMERICAN TROTTING-HORSE.

II. S. Randall, in the introduction to Youatt on the Horse, in calling our attention to the American trotting-horses, says, "Though in reality they do not, at least as a whole, constitute a breed, or even a distinct variety or family, there is a family of superior trotters, including several of the best our country has ever produced, the descendants of Abdallah and Messenger, and running back through their sire, Mambrino, to the thoroughbred horse Old Messenger. But many of our best trotters have no known pedigrees, and some of them, without doubt, are entirely destitute of the blood of the race-horse. Lady Suffolk is by Engineer, but the blood of Engineer is unknown. (She is a gray mare, fifteen hands and two inches high.) The Dutchman has no known pedigree. Other celebrated trotters stand in the same category, though we are inclined to think that a decided majority of the best, especially at long distances, have a greater or less infusion of the blood of the race-horse."

The United States has undoubtedly produced more superior trotters than any other country in the world, and in no other country has the speed of the best American trotters been equaled. The "Spirit of the Times," published in New York, contains lists of hundreds of matches. A number of these performances will be selected, enough to show that the excellence of American trotters is not founded on a solitary achievement or very rare cases, nor to be ascribed to the possession of any distinct and peculiar breed of horses, but it is the natural and common fruit of that union of blood and bone which forms the desideratum in a good hunter, with the addition of skillful training and artful jockeying in the trotter.

"Nimrod," in admitting the superiority of our trotting-horses to the English, claims that the English approach very near to the Americans, even in this breed of cattle. It is possible that his characteristic national vanity would not allow him to make a further confession. But Mr. Youatt says that there is no comparison whatever between the trotting-horses of the two countries. Mr. Wheelan, who took Rattler to England, and doubly distanced with ease every horse that ventured to start against him, as the record shows, informs us that there are twenty or more roadsters in common use in this city, that could compete successfully with the fastest trotters on the English turf. They neither understand the art of training, driving, or riding, there.

For example: Some few years since, *Alexander* was purchased by Messrs. C. & B., of the city of New York, for a friend in England. Alexander was a well-known roadster here, and was purchased to order at a low rate. The horse was sent out, and trials made of him, but so unsuccessful were they, that the importers considered him an imposition. Thus the matter stood for a year or more. When Wheelan arrived in England, he recognized the horse, and learned the particulars of his purchase and subsequent trials there. By his advice the horse was nominated in a stake at Manchester, with four or five of the best trotters in England, Wheelan engaging to train and ride him.

When the horses came upon the ground the odds were four and five to one against Alexander, who won by nearly a *quarter of a mile!* Wheelan says he took the track in starting, and widened the gap at his ease; that near the finish, being surprised that no horse was anywhere near him, as his own had not yet made a stroke, he got frightened, thinking some one might outbrush him; that he put Alexander up to his work, and finally won by an immense way, no horse getting to the head of the quarter stretch as he came out at the winning stand. The importers of Alexander, at any rate, were so surprised and delighted at his performance, that they presented him (Wheelan) with a magnificent gold timing-watch, and other valuable presents, and sent Messrs. C. & B. a superb service of plate, which may be seen at any time at their establishment in Maiden Lane.

Here it is clearly shown that the comparative speed of the American horses is to be attributed not to *breed*, but to *management*, on which we the rather insist, as it is to be desired that American agriculturists, and all breeders and trainers of horses, should understand and practice upon some fixed and rational principles, rather than rely for success on some imaginary strain of horses of no certain origin or established blood. After all, we have accounts of performances in trotting, by English horses, that may be considered as extraordinary as those of our own, when allowance is made for the greater value placed, and the more attention and skill bestowed, upon trotters in this country than in that.

The celebrated English trotter Archer, descended from old Shields, a remarkably strong horse, and master of fifteen stone (two hundred and ten pounds), trotted his sixteen miles in a very severe frost in less than fifty minutes. In 1791, a brown mare trotted, in England, on the Essex road, sixteen miles in fifty-eight minutes and some seconds, being then eighteen years old; and a mare by the name of Lady Hampton, on the 2d day of May, 1842, in England, trotted seventeen miles in fifty-eight minutes and thirty-seven seconds in harness. She was driven by Burke, of great English trotting celebrity. On the 13th of October, 1819, a trotting match was decided over Sunbury Common, between Mr. Dixon's brown gelding and Mr. Bishop's grey gelding, carrying twelve stone (one hundred and sixty-eight pounds) each, which was won by the former in twenty-seven minutes and ten seconds. A Mr. Stevens made a bet, which was decided on the 5th of October, 1796, that he would produce a pair of horses, his own property, that should trot in tandem from Windsor to Hampton Court, a distance of sixteen miles, within the hour; notwithstanding the cross-country road, and the great number of turnings, they performed it with ease in fifty-seven minutes and thirteen seconds. The Phenomena mare trotted nineteen miles in an hour.

In order to perform these great feats of speed it needs the combination of form and blood with bone and muscle, which gives distinction to the hunter, and muscle and wind to the trotting horse. Many a celebrated trotting horse is put to his work too young, and thereby his usefulness destroyed. Hiram Woodruff, and there can be no better authority, would commence a horse's training for the trot at five or six years of age, giving him light work however, but going on increasing up to nine or ten years, and with kind usage he might continue up to this mark for three or four years longer, and they often last to perform admirably until after nineteen. The stoutest horses, of whatever kind of blood, will give way if put at training under heavy weights, for trotting stake or chase; but on the other hand, without blood to give him wind and courage, what would avail his mass of flesh and bone in a trial to trot his hundred miles in ten hours!

Johnson, author of the Sportsman's Cyclopedia, justly esteemed high authority on the subjects, remarks that " thoroughbred horses, and particularly those of the best blood, are seldom possessed of sufficient bone to render them pre-eminently calculated for the chase; yet I am free to confess that the very best hunters and trotters that have fallen under my observation have been remarkably well and very high bred, but yet none absolutely thorough-bred." The cases of Abdallah and Messenger have been instances to show that great trotters not thorough-bred, may and do *beget* trotters, and hence some would argue that a distinct race of horses may or does exist. But it is to be remembered that both Abdallah and Messenger are sons of Mambrino, son of Old Messenger and of a Messenger mare, though not thorough-bred; and nothing is better known by all who have been in the habit of attending to these subjects than that the Messenger family is distinguished for making first-rate coach horses—quick in light harness, and remarkable for endurance and long life. That Abdallah, therefore, himself deep in the Messenger blood, should be himself a trotter and a getter of trotters, only proves that like begets like, and that of a distinct breed like the thorough-bred horse, characterized by the possession of general properties belonging only to and constituting that breed, there may be a *peculiar* family of the same breed. Thus we have the three classes of the English thorough-bred stock, to wit: the *Herod*, the *Matchem*, and the *Eclipse*, that have served as crosses for each other. A dash of the blood of Old Messenger imparts high form and action for the stage-coach, and the eye of the connoisseur can detect the signs in a horse in whose veins flow even one eighth of his blood. So the fact is generally known to old gentlemen in the South, and especially avouched by sporting and agricultural men in South Carolina, that the stock of old Jaus (there called Genius), were so remarkable as *road* and *saddle horses* as to have gotten to be considered a distinct breed. So the Topgallant stock made fine saddle-horses, excelling in the canter. The blood-horse, too, is remarkable for longevity—the Messenger stock particularly so. If the truth could be known, it is probably in large or smaller streams in each of the four thorough-breds which the late General Hampton drove in his coach all together for sixteen years.

Here are some extracts from a letter from Col. N. Goldsborough, of Maryland, who has an eye for the fine points of a horse, and who has thought much on the subject. He, in confirmation of our hypothesis, says, speaking of Tom Thumb:

"By whence came his lastiness, his powers of endurance, as well as speed? I have been in the habit of thinking that no horse could long continue exertion, especially at a rapid pace, without a good tincture of the blood."

"About the same time there went to England a horse called Rattler, of great speed as a trotter; he was represented as a cross of a full-bred horse on the Canadian mare. What a magnificent picture 'Whalebone' makes in his trotting action, and how different from the above-named horses! Philip Hemsly, of Queen Ann's County, England, was said to own a famous mare, much in style of the famous Phenomena mare of England, which could keep up with a pack of hounds all day in a trot; and it is said that she could pass over the largest oak bodies lying in the wood, without breaking up. I have been informed that some of the best trotters that ever were in New Jersey, were the offspring of Monmouth Eclipse—the Messenger blood, you see! I know of no other family of the pure blood horse which may be said emphatically to produce *trotters* —the exception confirms the rule. Col. Lloyd's Vingtun and old Topgallant got fine *trotting* and racking horses. Is there more than one out of twenty thorough-breds that makes really a *racer?* And are there not as many trotters at the North— and more—than there are racers at the South, &c., where the most systematic efforts have been persevered in for years, exclusively for the production of racers? I have often wondered whence they of the North derive their horses. From what I have seen and heard, they have a peculiar family, different in appearance, in form, strikingly from ours. They of the North have had some method in this matter, as well as the breeders of short-horned Leicester sheep, &c. About the Lakes they have a horse of great speed and power, as I am informed, called the Frencher. The English officers bring over from the mother country fine-blooded stallions for troopers and parade. It is the cross of these and the Canadian mares which produces the Frencher. *Blood* is indispensable. The Canadian is the *unde renit*. They are descended from the horses of Normandy, carried over by French settlers. Napoleon's coach, when captured, was being drawn by four *Norman horses*. It would appear by this, that the Emperor was not fond of sitting behind *slow cattle*. When the Spaniards were in possession of the Low Countries, they carried with them their Andalusian horses; these were crossed on the Normans, which produced great improvement. When the Spaniards were expelled, the breeding-in from this stock must have produced a distinct family, as Bakwell produced with our race quadrupeds. Climate necessarily produced a change in the Norman horse, when transferred to the rigor of Canadian winters—hence the thick coat of hair, &c. The Andalusian, you know, is of Arabian descent. So far as I have been able to learn, Vermont is indebted to Canada for her distinguished race of roadsters, as well as the neighboring States. They have one distinguished family—the "Morgan"—descended from a little Canadian, famous, too, for running quarter races. This family has been cherished for years, and is as distinguished among them as old Archer was in Virginia."

It is now in proof that this Morgan breed is descended from a horse that was stolen from General De Lancey, importer of Wildair; and there is every reason to believe that though he may not have been a thorough-bred, he was well steeped in the best blood of the Anglo-American turf-horse.

While it has been found impracticable to obtain any precise information as to the pedigree of some of our very best trotters, in other cases, where more is known, they are found to be deep in the blood. Awful, is known to have been gotten by a thorough-bred, "American Boy."

Lady Suffolk is by Engineer, not known. Abdallah, as before mentioned, is by Mambrino, and he again, a great trotter, by Messenger; but Dutchman, one of our best trotters, has no known pedigree, though we have some reason to think he was by Young Oscar, then at Carlisle. He was taken out of a clay-yard, and was transferred to the trotting-turf from a Pennsylvania team-wagon. The Canadian or Norman-French stallions, small and compact, which, on well formed, large mares, give such fine harness horses and trotters, are, as before said, deeply imbued with the blood of the Barb, taken from Spain into Normandy. The infusion of blood into our coach horses has enabled them to lengthen their stages, and in very observable proportion to the degree of blood.

Finally, as where the blood of the trotter, when known, is seen to flow in so many instances from a spring of pure blood, is it not fair to infer a similar origin in cases where the blood cannot be traced? especially as the universal experience of all times proves that in other places the cases have been *exceedingly rare* in which a horse of impure blood may be a great trotter at long distances, because his speed at his best is greatly behind that of our best speed on the turf. But it would, according to all principles of reasoning, be unreasonable to expect great excellence, even as a trotter, in horses *altogether* free from blood, which gives foot and wind to the Eastern courser. Though we may not be able to trace it, and though in solitary cases a horse without it may possess great speed and lastiness in the trot, from excellent accidental conformation, we repeat that the possession of the two warrants the presumption of the third, however obscure the trace or remote the origin: this is our theory. But the action to be cultivated in the racer and trotter, is of itself sufficient to explain why a racer should not succeed at once on the turf and on the trotting course. All reflecting and observing men will admit that there is but one way for a horse to excel in any particular business. Whatever that business may be, to be perfect in it, he should be educated and kept to it, and to it only. *A trotting horse should do nothing but trot.*

EXTRAORDINARY PERFORMANCES OF AMERICAN TROTTERS.

On the 21st of October, 1841, a match came off on the Centerville Course, L. I., between *Americus*, a bay gelding, and *Lady Suffolk;* five-mile heats, purse $3,500. Americus won in two heats, and in the following time:

FIRST HEAT.		SECOND HEAT.	
Time of first mile,	2.44½	Time of first mile,	2.51
" second "	2.50½	" second "	2.50
" third "	2.46	" third "	2.46
" fourth "	2.42½	" fourth "	2.47
" fifth "	2.44½	" fifth "	2.44½
Time of first heat,	13.58	Time of second heat,	13.58½

Topgallant, by Hambletonian, trotted in harness 12 miles in 38 minutes. (See Turf Register, Vol. I., p. 124.)

Ten miles have been repeatedly trotted in America within two or three seconds of thirty minutes.

A roan mare, called *Yankee Sal*, trotted, as has been stated without contradiction, in a match against time, on the course at Providence, R. I., which was at the time heavy and deep, fifteen miles and a half in 48 minutes and 43 seconds, a rate of speed so prodigious under the circumstances, that we have often suspected there may have been an error as to the time.

Lady Kate, a bay mare, 15 hands high, trotted on the Canton Course, near Baltimore, 16 miles in 53 minutes and 13 seconds; and, the reporter adds, "she could have done seventeen with ease."

In October, 1831, *Jerry* performed 17 miles, on the Centerville Course, L. I., in 58 minutes, under the saddle.

In September, 1839, *Tom Thumb*, an American horse, was driven, in England, 16½ miles in 58 minutes. We shall have more to say of this phenomenon when we come to his 100 miles performance.

In 1856, the gray gelding *Mount Holly* was backed, at $1,000 to $500, to trot twenty miles within the hour. The attempt was made on the 10th of October, on the Hunting Park Course, Pa., but failed. He performed 17 miles in 53 minutes and 18 seconds, without the least distress. He was miserably jockeyed for the first five miles, doing none of them in less than five minutes.

Pelham, a large bay gelding, in a match to go 16 miles within the hour, performed that distance without any training, in 58 minutes and 28 seconds. He went in harness seven miles in 26 minutes and 29 seconds, when, the sulky being badly constructed, he was taken out and saddled, and mounted by Wallace (weighing 160 lbs. without his saddle), and won his match.

Paul Pry, a bay gelding, was backed to perform 17¾ miles within the hour, under the saddle, on the 9th of November, 1833, on the Union Course, L. I. He won the match, performing 18 miles in 58 minutes and 53 seconds. Hiram Woodruff, weighing then 138 lbs., jockeyed him. Paul Pry was nine years old, bred on Long Island, and got by Mount Holly, dam by Hambleton.

In 1831, *Chancellor*, a gray gelding, ridden by a small boy, performed 32 miles on the Hunting Park Course, Pa., in 1 hour, 58 minutes and 31 seconds. The last mile, to save a bet, was trotted in 3 minutes and 7 seconds.

In October, of the same year, George Woodruff drove *Whalebone*, on the same course, the same distance in 1 hour, 58 minutes and 5 seconds. He commenced the match in a light sulky, which broke down on the fourteenth mile, and was replaced by one much heavier. This course is fifty feet more than a mile in the saddle track, and much more than that in the harness track.

On the 11th of September, 1839, Mr. McMahon's bay mare, *Empress*, on the Bacon Course, in a match against time, $600 a side, performed, in harness, 33 miles in 1 hour, 58 minutes and 55 seconds.

The American horse *Rattler* was ridden by Mr. Osbaldistone, in England, in a match against *Driver*, 34 miles, in 2 hours, 18 minutes, and 56 seconds. Mr. Osbaldistone weighed 125 lbs. Rattler was 15½ hands high.

In July, 1835, *Black Joke* was driven, in a match against time, on the course at Providence, R. I., 50 miles, in 3 hours and 57 minutes.

A gray roadster is reported to have performed about the same distance on the Hunting Park Course, Pa., in 3 hours 40 minutes. It was a private match.

A gray mare, *Mischief*, by Mount Holly, out of a Messenger mare, 8 years old, in July, 1837, performed about 84¼ miles in 8 hours and 30 minutes, in harness, on the road from Jersey City to Philadelphia.

Tom Thumb, before mentioned, on 2d of February, 1829, on Sunbury Common, England, performed 100 miles in 10 hours 7 minutes, in harness. He was driven by William Haggerty (weighing over 140 lbs.), in a match cart weighing 108 lbs. This performance, so extraordinary, demands more than a passing notice, and we accordingly abridge from an English paper the following description:—

"Tom Thumb was brought from beyond the Missouri, and is reported to have been an Indian pony, caught wild, and tamed. Others again, allowing him to have been thus domesticated, think him to have been not the full-bred wild horse of the Western prairies, but to have had some cross of higher and purer blood. But too little is known of his breeding, save his Western origin, to justify any satisfactory speculation.

"His height was 14¼ hands, and his appearance, when standing, rough and uncouth. From his birth he had never been shorn of a hair. He was an animal of remarkable hardihood, a hearty feeder, and, though accustomed to rough usage, was free from vice, playful, and good-tempered. He was eleven years old when he performed his match, and had never a day's illness. At full speed his action was particularly beautiful, he threw his fore-legs well out, and brought up his quarters in good style; he trotted square, though rather wide behind, and low. He was driven without a bearing-rein, which is going out of use in England, and simply with a snaffle-bit and martingale. He pulled extremely hard, his head being, in consequence, well up and close to his neck, and his mouth wide open. He did his work with great ease to himself, and, at 11 miles the hour, seemed to be only playing, while horses accompanying labored hard. The whole time allowed for refreshment during his great performance amounted to but 35 minutes, including taking out and putting to the cart, taking off and putting on the harness, feeding, rubbing down, and stalling. The day before and the day after the match, he walked full twenty miles. His jockey provided himself with a whip, but made no use of it in driving him, as a slight kick on the hind quarters was quite sufficient to increase his speed when necessary."

In February, 1828, a pair of horses trotted against time, 100 miles, on the Jamaica Turnpike, on Long Island, and won in 11 hours and 54 minutes.

In June, 1834, a pair of horses belonging to Mr. Theal, trotted that distance, in harness, on the Centerville Course, within ten hours; and immediately after, Mr. B—— offered to bet $5,500 that he would produce a pair of horses that could could trot 110 miles in harness within the same time! The bet was taken, and afterwards abandoned by the backers of time, who paid forfeit. Another gentleman offered to produce, for a wager, a pair of horses that should trot 100 miles in nine hours in harness, but no one would back time against the performance.

Having gone through with these numerous details, let us dwell for a moment on the remarkable performances of the Dutchman, which are taken from the "Spirit of the Times." A trotting match for $1,000 a side; three-mile heats, under saddle, came off on Saturday, Oct. 6th, 1838, at 4 o'clock, over the Bacon Course. The time made was, at this time, by far the best on record.

Dutchman and *Rattler* were the contending horses; the first was a handsome bay gelding, of great size and substance, about sixteen hands high; he is what is termed a "meaty horse," and looks, when in fine condition, like an ordinary roadster in "good order." He was trained for the match and and ridden by Hiram Woodruff. Rattler was a brown gelding, of about 15½ hands, a "rum 'un to look at;" he was drawn very fine, though one of those that seldom carry an ounce of superfluous flesh; his feed seldom exceeded six quarts per day, while the Dutchman's had been between twelve and sixteen. Rattler was trained and ridden by William Wheelan. His style of going is superior to Dutchman's; he spreads himself well, and strikes out clear and even. Dutchman does not appear to have perfect command of his hind legs; instead of throwing them forward, he raises them so high as to throw up his rump, and consequently falls short in his stride. The main dependence of his backers was based upon his game; and a gentleman who bet heavy on Rattler, offered to 2 to 1 on the Dutchman before the start if the heats were broken.

The odds, before the horses came upon the track, were 5 to 4 on Dutchman; after the riders were up, 5 to 3 was current, and at length to 2 to 1. As they were ridden up and down in front of the stand previous to starting, both appeared to be in superb condition, and to have their action perfectly. The track was so hard and smooth that the nails in the shoes of the horses could be seen every step they made. A great many bets were made on time; even bets were made that it would be better than any on record. To determine what the best time on record was, it was shown that in 1833, *Columbus* trotted a three-mile heat, under the saddle, over the Hunting Park Course, Philadelphia, in 7 minutes and 57 seconds, which was declared to be the best time made. On the 10th of October, 1837, Daniel D. Tompkins, in a match, beat Rattler over the Centerville Course, in 7.59, 8.09, three-mile heats, under the saddle. Both Dutchman and the Rattler were at this time owned by a gentleman in New York city.

THE RACE.—Rattler drew the track, but resigned it to Dutchman on the first quarter; he came in front on the back-side, and at the half-mile post led by two lengths; he soon after broke up, when Dutchman headed him, and led him past the stand (2.42), round the straight stretch on the back-side, when, the ground being descending, and more favorable to him, Rattler passed. Dutchman waited upon him, close up, to near the three-quarter-mile post, where Rattler shook him off, and led past the stand, in 2.38, by four lengths; keeping up his rate, he led down the back-side, and round the turn to the straight stretch in front, when Hiram caught him by the head, and laid in the spurs to the gaffs; the brush home was tremendous, but Rattler won by nearly a length, trotting the third mile in 2.37, and the heat in 7.54½.

Second Heat.—Dutchman broke at starting, and two to one were offered against him. Down the back-side the horses were lapped all the way; on the ascending ground, within about ten rods of the half-mile post, Dutchman gained a little, and came first to the stand in 2.37. He drew out two lengths ahead round the first turn on the second mile, but Rattler gallantly challenged him down the back-side, and lapped him. At the half-mile post Dutchman was again clear, but by a desperate effort Rattler lapped him when they got into straight work in front, and thus they came down to the stand, in 2.33. On the back-side, Rattler, as usual, drew out clear, but for an instant only; the spurs were well laid into Dutchman, and the struggle was desperate. Dutchman hung down upon Rattler's quarters, and gradually gained to the half-mile post, when they were locked as perfectly as if in double harness. The contest was almost too much for Rattler, who skipped several times, and was only prevented from breaking by Bill's holding him up. They came up the quarter-stretch at an immense pace, but opposite the four-mile distance-stand Rattler unfortunately broke up, when nearly a length ahead, and Dutchman won the heat by nearly eight lengths. When Rattler skipped, Wheelan should have taken him in hand; but he was so much ahead, and so near home (within 180 yards), that under the intense excitement of the moment he neglected doing so; had he done so, however, at the rate Dutchman was going, he would probably have won by a few feet, for Rattler could not have made up any leeway caused by pulling him up; nothing but his breaking lost him the heat. The instant Rattler broke, Hiram pulled up Dutchman, and he would have walked out, had not the people on the instant cried out, "Come on!" The last mile was performed in 2.40, and the heat in 7.50; had Dutchman kept up his stroke, the time of the heat would have been 7.48.

Third Heat.—Dutchman went off with a fine stride (two to one offered on him), and led about half-way down the back-side, when Rattler caught him; at the half-mile post they were locked, and thus they came to the stand, in 2.42. They made the turn in the same position, and nothing but repeated injunctions from the judges to keep silent prevented cheers from the stand that would have made the welkin ring. It was a beautiful sight. Both were going, Dutchman under the spur, at a great flight of speed, neck and neck. Half-way down the back-side, Rattler got almost clear, but Dutchman soon after lapped, and when they came to the stand (2.38½), was half a length ahead. When they got into straight work on the back-side, Rattler again collared him, and they went locked to nearly the half-mile post, when Dutchman once more got in front, Wheelan having taken Rattler in hand for a brush up the straight side. This he made soon after. They were lapped as they swung round the turn, and the struggle that now ensued revived recollections of Bascomb and Postboy. Profound silence was preserved on the stand, that neither horse might be excited or frightened into a break, and the interest of the scene was so great that each of the spectators seemed to hold his breath as the horses neared the stand. It was a brush to the end, Dutchman coming out a throatlatch in front, caused by Hiram's giving up his pull, and giving him a push, which made him clear the winner by a foot. The excited feelings of the crowd in the stand could no longer be repressed, but burst out in a tumultuous cheer, that might have been heard three miles off. The last mile was done in 2.41½, and the heat in 8.02. The judges, after some discussion, pronounced it a dead heat.

Great odds were now offered on Dutchman, though he exhibited more "signals of distress" than Rattler; his trainer, however, informed us that he "hung out" these after taking his ordinary exercise; it was a way he had, rather than any severe exertion which produced them. Both sweat freely, and came to the post a fourth time "about as good as new." The performance of the match commenced at four o'clock; it was six, and almost dark, when they started on the

Fourth Heat.—Dutchman led off from the score to halfway down the back-side, by three lengths; Rattler, however, lapped him at the half-mile post, but the Dutchman soon after drew out in front again. Hiram kept him at his work from this point to the finish, and Rattler never got up to him afterwards, that we could see, for it was now so dark that neither horse could be distinguished; Rattler subsequently fell off in his stride, and was finally beaten handily by six lengths, after as honest a race as we ever saw, and by far the best time on record at this date.

From the above it will be seen that the average time of the second heat was 2.36 and two thirds of a second per mile; and that of the four heats 2.40 and five sixths of a second.

A great number of people were assembled to witness the match, and we were struck with the number of gentlemen in attendance. Every one seemed delighted, and as they will no doubt be induced to turn out on any similar occasion, the match cannot fail to exercise a salutary and beneficial influence upon our Associations for the Improvement of Road Horses. In closing our account, we must not omit to speak of the admirable condition in which Woodruff and Wheelan brought their horses to the post; they were jockeyed most admirably, displaying the most consummate skill and judgment; a superior exhibition of horsemanship has not been seen since the day Purdy stripped to throw a leg over the saddle of Old Eclipse.

From the same paper we copy a report of a match against time, which came off in the following year, 1839.

This match, for $1,000 a side, vs. time, was made on the 11th July, on the evening of the day on which Dutchman beat Awful, three-mile heats, in harness, in a match of $5,000 vs. $2,500. The backers of time staked their money against Dutchman's trotting three miles in 7.49. He was allowed to perform the match in harness or under the saddle, to make two trials if necessary, and to have two hours' intermission between them; the match was appointed to come off on the first day of August, provided the weather and track were unexceptionable; weight according to the rules of the course, or 145 lbs.

Fortunately the track was in pretty good order, though dusty; the weather all day had been excessively warm, but as the match came off rather late in the afternoon, the air was cooler and more bracing. After being walked for some time up and down in front of the stand in his match-cart, with his hood and sheet on, he was taken out of his harness and groomed; at a quarter to seven o'clock was led to the judges' stand, and Hiram Woodruff coming out of the weighing-room, threw his leg over the saddle. A fine thorough-bred mare was also mounted by Isaac Woodruff at the same time, to keep him company, and at a steady racing pace. The judge and the two official timers now selected a third, who having taken his place in the stand, the horses were called up. Dutchman was the favorite at odds.

THE RACE.—At precisely ten minutes to 7 o'clock the signal was given, and Dutchman went off with a long, clean stroke, that kept the mare up to three parts racing speed; Dutchman went to the quarter-mile post in 40 seconds, and did the first half mile in 1.17½; the mare was not allowed to pass him, but was kept well up ; in coming down the quarter-stretch Dutchman pulled to the mare, doing the first mile in 2.34½, At the stand Hiram told her rider to "*go along*," and as she locked him, old Dutchman made a tremendous burst, doing the first quarter of the second mile in 38 seconds, and the half mile in 1.15. Going down the back-side Hiram bade Isaac "let the mare out," and so immense was Dutchman's rate for a few hundred yards, that it seemed as if the mare could not have passed had she tried. From the half-mile post to the stand there was no faltering, and but little falling off in the pace, the mile being done in 2.28—the best time on record. Dutchman was kept at his work from the stand, and came to the quarter-mile post on the third mile in 39 seconds, and to the half-mile post in 1.16, which showed a falling off but of a second from the time of the previous first quarter and first half-mile. Hiram feeling confident now that he had won the match, and all the bets against time, came home at an easier pace, finishing the third and last mile in 7.32½, being sixteen and a half seconds inside of time.

Dutchman, in this match, made the best time on record, at one, two, and three miles, and I believe that it has never been beaten at three miles. He was in superb condition, and never broke up from the start to the end. We need hardly add, he was jockeyed most admirably.

ETHAN ALLEN.

A brief synopsis of the pedigree and performances of this famous trotting-stallion, compiled from the columns of "Wilkes' Spirit of the Times," will not be out of place here. This noted little trotting-horse does not stand above 14½. He was foaled at Ticonderoga, in this State, in 1849, on the farm of Mr. Holcomb, one of his present owners. There is a difference of opinion as to his sire; some contend that he was sired by Vermont Black-Hawk, some by Flying Morgan ; but there is nothing decisive known with regard to it. His dam was an ordinary road-mare, who was sold at fifteen dollars. Ethan gave early promise of being a great trotter, and his performances have been numerous and varied. His first, was at the Clinton County Fair, held at Keeseville, N. Y., for a purse of fifteen dollars, given for three-year-old stallions, at mile heats, in harness. He won, beating two competitors, over a heavy track, in 3.20 and 3.21. He appeared, in May, 1853, on the Union Course, L. I., as a four-year-old, against Rose of Washington, and he won in three straight heats, in 2.26, 2.39, 2.42. In the fall of 1855, he beat North Horse and Stockbridge Chief. In 1856, he overcame the stallion Hiram Drew; and, flushed with these victories, in November of the same year, he ventured to meet Flora Temple at Boston, but met from her a defeat in 2.32 and 2.36. Last year he ventured for the stallion-purse given by the Union Association, at mile heats, 105 lb. wagons, and distanced George M. Patchen in 2.28, thus equaling the best time ever made. Shortly after this the two double-team matches with Ethan and Mate and Lantern and Mate came off, resulting in the victory of Ethan, thus placing him second in renown to Flora Temple. At this point the proprietor of the Fashion Course offered a purse of $2,000, and Flora and Ethan came together on May 31, 1859, mile heats, best three in five, to wagons. The race created great excitement, and over fifteen thousand persons were present to witness it. It was won by Flora, in three straight heats, in 2.25, 2.27, 227. He then went against Tacony for a purse of $1,000, at Burlington, Vermont, and beat him in three straight heats, in 2.37, 2.36, 2.33. He again came against Flora Temple, November 24th, of that year, for a purse of $1,500, mile heats in harness, best three in five. He won the first heat in 2.27, but lost the three in succession in 2.26, 2.27, and 2.25½. He had thus established his fame by his works, and was held to be the Champion Trotting-Stallion on the American turf—but has since been surpassed by the world-famed trotting-stallion George M. Patchen, as will be seen by the great trot of the Champion Stallions on Long Island, which came off over the Union Course on Wednesday, May 16th, 1860,—purse $2,000 each.

The sky was overcast with thin clouds, one great veil, just thick enough to screen all faces from the glare of the sun, and no more. The air was balmy and invigorating, and a pleasant breeze ruffled among the tender May blossoms of the road-side orchards, and fluttered the rich young foliage of the shrubberies, as we swept down the vistas between them. About the hotels near the Course, thousands were congregated when we reached there, although we were there early; and the constant patter and rush of speedy animals gave token of a brilliant and immense gathering. After a brief pause here, we proceeded to the

Course, where some ten thousand spectators soon after gathered to witness the trot, mile heats in harness, best three in five, between the famous stallions Ethan Allen and George M. Patchen.

First Heat.—Soon after 3 o'clock the horses were called up Ethan having drawn the pole. In scoring he outfooted Patchen the first time, and they did not go off. At the next attempt, however, a capital start was effected, and away they went at a great pace. In rounding the turn Ethan drew ahead about a length, and thus they went to the quarter pole in 37 seconds. The half mile was reached in 1.12, and Ethan was still about a length ahead. He was unable to increase his speed, and Patchen drew upon him slowly but surely, and they swung into the home-stretch lapped. Now it was nip and tuck, and the rate of going tremendous. Half-way up the stretch Ethan broke, but Pfifer brought him down again in a masterly style, and he seemed to lose nothing. They came thundering on as near head and head as might be, but at crossing the score Patchen was a short nick ahead. The time was 2.25; and the wind now too keen and rigid for the horses to open well.

The trotting in this heat was superb, and only the great length of his stroke enabled Patchen to get his head in front of the little gentleman. The former never made a skip. The result struck the backers of Ethan all of a heap, and the odds veered round like a weather-vane in the "variables." Big odds could have been had on Patchen, but takers were as scarce as they had been before. All the time-bets were decided by this first heat, nobody having marked below 2.25.

Second Heat.—In scoring the first time for this heat Ethan broke, and they did not get the word. Next, Patchen was out-footed, and there was no go. Then they came up square, but Patchen broke upon the score, both at a great pace, and it was "no go." At the fourth attempt they got off very evenly, and the pace was tremendous. Again the great, steady stroke of Patchen told, and after rounding the turn he was a length ahead. The quarter-pole was reached in 36 seconds, and a half mile in 1.11, Patchen leading a length and a half. Swinging into the home-stretch Patchen still led, and half-way up Ethan broke. The former won the heat, by three lengths in 2.24. This was a clipper for entire horses, and Patchen had a burst of cheers. As he stood under his blankets, after being brought back

GEORGE M. PATCHEN

to the stand, we noticed that he was playing with his grooms—a mighty good sign after two severe heats. The wind was still raw and cold, but both horses sweated out well between this heat and the next.

Third and last Heat.—There were now a little too many ineffectual attempts at scoring. The first time Patchen dropped behind; at the next he broke upon the score, and this was the case again; another try, and he again lagged; but at the fifth attempt they got the word, going finely together. They were neck and neck round the turn, but at the quarter, which was reached in 37 seconds, Patchen had a lead of nearly a length. At the half mile, which was passed in 1.13, he had increased his lead to two lengths. In trying to get up with this great rate of going and commanding stride, Ethan now left his foot and took a little run. He went on gamely after his opponent when he caught again, but was never able to reach him; and the redoubtable George M. Patchen crossed the score three open lengths ahead, and apparently with a link or two not let out. The time of this heat was 2.29.

The winning of this race by Patchen in three straight heats, and in the time we have above recorded, has put him into the very front rank. He did not break in either of the heats, and though the track was in admirable order, there was "a nipping and an eager air,"—one not calculated for very free action, and the fastest time. Patchen has risen vastly in the estimation of his friends, and some of them go so far as to think him the "coming horse," to dethrone the imperial Queen of the trotting turf, Flora.

For Ethan Allen, we cannot say but that he trotted in better time than any outsider supposed he could have done. It is not that he has fallen off, but that Patchen has far surpassed the opinion commonly entertained of his powers.

GEORGE M. PATCHEN.

We now come to George M. Patchen, the fastest trotting-stallion in the world, that we have any account of. He is a handsome bay horse, of the Bashaw blood, stands fifteen hands and three inches high; he was got by Cassius M. Clay, by Henry Clay, by Andrew Jackson, by Bashaw. Patchen is out of a dam of whose pedigree nothing is known, and he was foaled on Long Island. He was purchased by Messrs. Longstreet & Buckley, when young, and kept by them as a covering stallion, at Bordentown, New Jersey, up to the time of his *debut* upon the turf with Ethan Allen, in 1858. Patchen achieved no laurels in his first essays, for the little gentleman distanced him the first heat, in 2.28. His friends, however, did not lose confidence in him, but entered him for a purse of $1,000, given by the proprietor of the Fashion Course, on the 9th of June, 1859. Brown Dick, Miller's Damsel, and Lady Woodruff were entered. Brown Dick won in three straight heats. On the 21st of the same month, Patchen went against Lady Woodruff and Brown Dick, on the Union Course, L. I., mile heats, best three in five, for a purse and stake of $800. Lady Woodruff won. The next week he met Lady and Brown Dick again, over the same course, two-mile heats, and won in 5.01 and 5.03. He beat Lady Woodruff two days after, in a match for $1,000 a side, mile heats, best three in five, to wagons and drivers of 300 pounds. The shortest time in these heats was 2.30. July 7th, he beat Miller's Damsel and Brown Dick on the Union Course, best three in five. Five heats were trotted in 2.26, 2.26, 2.29, 2.28, and 2.29. Brown Dick, however, turned the tables upon him in Philadelphia, when he beat him three straight heats, in 2.31, 2.29, and 2.30. Two days afterwards Patchen's star was again in the ascendant, he having beaten Brown Dick and Lantern over the Point Breeze Course. The best time in this race was 2.29. He then returned to New York, and went against Flora Temple, on the 21st of November, on the Union Course, L. I., for the proprietor's purse of $1,000, Flora in harness, the stallion under saddle. The mare won the first and second heat, but broke and swerved across the horse in the third. This heat was given to Patchen by the judges. The time was 2.24. All further proceedings were postponed in consequence of darkness having come on, and no more heats were ever trotted for the money, a compromise being affected between the interested parties. This is the last race Patchen went prior to that we are now reporting. A loud hum of expectation had arisen, and there was pulling-out of watches among the thousands who were conversing upon the quarter-stretch, when the crowd split asunder, as the waters of the Red Sea parted before the prancing chargers of the Israelites. It was to let Patchen through. Between two grooms, who could scarcely retain his eagerness, he came, a type of equine pride and power. Nor was his gallant little competitor far behind. He also came through the thick of the throng, with sprightly and elastic tread, and attended by a brace of grooms. The two horses reminded us of the Benicia Boy and Tom Sayers, and, as in that great tournament, the odds was upon the little one. Offers had been current of 100 to 60, in the morning, and as the time for the start drew nigh, two to one might have been had. Takers, however, were very shy, being deterred, we imagine, by the prestige of the little gentleman. Ethan Allen was the first to show, in a gay, green sulky, and driven by Dan Pfifer. His condition seemed to be all that anybody could desire. His coat was bright, his stride elastic and easy, and his way of going such as to confirm the confidence of his backers. He looked "all horse." Patchen then came along before D. Tallman, looking fit to trot for a man's life, and moving with a great elastic stride. As he came back to the stand after having been driven to the quarter-pole, we noticed that his broad nostrils had already began to expand, and we involuntarily exclaimed, "That horse will be very likely to upset the odds!"

First Heat—May 16th, 1860, on the Union Course, L. I. Soon after, the horses were called up, Ethan having drawn the pole. In scoring, he outfooted Patchen the first time, and they did not go off. At the next attempt, however, a capital start was effected, and away they went at a great pace. In rounding the turn, Ethan Allen drew ahead about a length, and thus they went to the quarter-pole in 37 seconds. The half mile was reached in 1.12, and Ethan was still ahead about a length. He was unable to increase his lead, while it appeared just about as plain that he would take a good deal of catching. Patchen, however, drew upon him slowly but surely, and they swung into the home-stretch lapped. Now it was nip and tuck, and the rate of going tremendous. Half-way up the stretch Ethan broke, but Pfifer brought him down again in masterly style, and he seemed to lose nothing. They came thundering on as near head and head as might be, but at crossing the score Patchen was a short nick in front. The time was 2.25, and the wind now too keen and rigid for the horses to open out well.

The trotting in this heat was superb, and only the great length of his stroke enabled Patchen to get his head in front of the little gentleman. The former never made a skip. The result

2*

struck the backers of Ethan Allen all of a heap; and the odds veered round like a weather-cock in the "variables." Big odds could have been had on Patchen, but takers were as scarce as they had been before. All the time-bets were decided by the first heat, nobody having marked below 2.25.

Second Heat.—In scoring the first time for this heat Ethan broke, and they did not get the word. Next, Patchen was outfooted, and there was "no go." At the fourth attempt they got off very evenly, and the pace tremendous. Again the great, steady stroke of Patchen told, and after rounding the turn he was a length ahead. The quarter-pole was reached in 36 seconds, and the half-mile in 1.11, Patchen leading a length and a half. Swinging into the home-stretch Patchen maintained his lead, and half-way up Ethan broke. The former won the heat, by three lengths, in 2.24.

This was a clipper for entire horses, and Patchen had a burst of cheers. As he stood under his blankets after being brought back to the stand, we noticed that he was playing with his grooms—a mighty good sign after two severe heats. The wind still continued raw and cold, but both horses sweated out well between this and the next.

Third and last Heat.—There were now a little too many ineffectual attempts at scoring. The first time Patchen dropped behind; at the next, he broke up on the score; and this was the case again. Another try, and he again lagged; but at the fifth attempt they got the word, going finely together. They were neck and neck round the turn, but at the quarter, which was reached in 37 seconds, Patchen had a lead of nearly a length. At the half-mile, which was passed in 1.13, he had increased his lead to nearly two lengths. In trying to get up with this great rate of going and commanding stride, Ethan now left his foot and took a little run. He went on gamely after his opponent when he caught again, but was never able to reach him, and the renowned George M. Patchen crossed the score three open lengths ahead, and apparently with a link or two not yet let out. The time of this heat was 2.29. The winning of this race by Patchen, in three straight heats, and the time we have above recorded, has put him in the very front rank. He did not break in either of the heats, and though the track was in admirable order, there was "a nipping and an eager air," not at all calculated for very free action and the fastest time. He has risen vastly in the estimation of his friends, and some of them go so far as to think him the "coming horse," to dethrone the imperial little queen of the trotting turf, Flora.

GREAT TROTTING MATCH ON LONG ISLAND.—*Flora Temple and George M. Patchen, the Emperor and Empress of the trotting turf.*—June 6th. Match for $1,000; mile heats, best 3 in 5, in harness. The day was clear and beautiful as could be desired, and thousands availed themselves of the opportunity of welcoming back to the turf the renowned little bay mare, after her winter of inglorious ease. Few trotting matches could have created a like excitement and general interest. Here on the one hand was the champion stallion backed by a host of friends; and here was the world-renowned queen of the trotting turf, whose performances may be found in the Life and Exploits of Flora Temple, as penned by Mr. Wilkes, and may be had at his office. Her friends had every confidence; they remembered her 2.19¾ at Kalamazoo, Michigan, and backed her heavily at the current rate of odds, 100 to 80.

Flora was driven by J. D. McMann, who gave her two or three smart brushes up the home-stretch to enliven her, and it had the desired effect. Patchen, who was driven by Darius Tallman, had the pole. The judges were Mr. Louis Beeral (the hero of the "Childers" letter in "London Field"), Mr. Parks, and Mr. J. Hammil.

First Heat.—After scoring five or six times, owing to Patchen outfooting Flora, they got the word, and started even. Immediately Flora broke, and Patchen opened a slight gap, but Flora, catching her trot, went quickly after him, and on rising the hill at the first turn she caught Patchen, and quickly changed position with him. Flora passed the quarter-pole a length ahead, in 35 seconds, and going down the back-stretch she increased the gap until near the half-mile pole; here Patchen began to close the gap, and they passed the half-mile in 1.11, Patchen again in front, and before reaching the turn he had opened a space of a length between them. Flora here again came to her work, and with a pretty show of speed glided upon him until they trotted side by side. At this point the race was very exciting. The crowd surged to and fro, and a murmur of intense interest ran through all, which was in creased to a wild and exciting cry as the gallant little bay, when within two lengths of the judges' stand, encouraged by the lash and loose rein, and the cheers of the anxious crowd, passed the stand a throat-latch in advance, in the extraordinary time of 2.21. No sooner had the judges recorded the time on the board than a deafening shout of admiration and wonder arose from the immense mass of people that surged to and fro below the judges' stand. Their favorite, the "little bay," had beaten even her own time on Long Island track, and they had witnessed the most magnificent contest that had ever been on that course.

Second Heat.—After one false start they got off even, and trotted steadily together to the turn. There Flora began to quicken her step, and again was the Jersey stallion destined to see her flying heels. She led past the quarter-pole by an open length. Patchen followed earnestly after, but as if seeing the fruitlessness of his efforts he took a fly in the air, and lost two lengths before he could be brought down. In going down the back stretch, he gathered together all his wondrous powers, and made after Flora, as if determined to win or die in the attempt, but the pace was too much for him, and he again lost his footing, as Flora passed the half mile two open lengths ahead, in 1.11¼. On the lower turn, the backers of Patchen had their hopes renewed for a moment by a plucky brush of Patchen, which closed the gap, and placed him nearly side by side with Flora. Up to the home-stretch they came home nip and tuck, neither having the advantage, amidst the shouts and cheers of the multitude. They neither heeded friend nor foe, but swept on, with the goal in sight. They strained every nerve and power; and Flora won by a neck in 2.24.

Third Heat.—Three false starts were made, and they were off for the third heat. They remained neck and neck until the first turn was made, when Flora took the lead, Patchen keeping close up. They passed the quarter-pole together, in 36 seconds. In going down the back-stretch Patchen closed and succeeded in passing Flora, and at the half-mile pole Patchen led; time 1.10. On the lower turn Patchen was an open length in advance. Flora now commenced to make her work, and quickening her trot as usual, she began to gain upon him. The contest was close and alternating until the distance-stand was reached, when Flora made one of her fearful efforts, which soon told

FLORA TEMPLE.

in her favor, Patchen taking a fly in the air, when within two lengths of the judges' stand, and Flora was hailed the conqueror by the united voice of the assembly. The time of this heat was 2.21½.

I shall say more of Patchen when I come to record Flora's time. I will only say that since this date Patchen has twice beaten Flora: once on the Union Course, Long Island, and once at Avolia Springs; and this is honor enough for one horse. If you want to know more about Patchen's or Flora's performances, purchase Wilkes' "American Racing and Trotting Record." In purchasing this book you have a complete calendar of all the trotting in the United States.

FLORA TEMPLE.

FLORA TEMPLE is a bay mare, 14 hands 1½ inches high, and weighs, in trotting condition, 835 pounds; was foaled in 1846, in Sangersfield, Oneida County, New York, out of Madam Temple, by One-eyed Hunter; he by the well-known Kentucky Hunter. In her younger days she was not at all valued for her speed, and was sold several times at a very low price. Those who are curious to know more about her history can procure her Life and Adventures, written by Wilkes, and for sale at his office. I shall record but few of her victories: only some of her best performances in 1859 and 1860,—for it would require a volume to record all of her races,—and these will be taken from Wilkes' Spirit of the Times.

At Chicago, Illinois, Sept. 16, 1859.—Purse $2,000; mile heats, best 3 in 5, in harness.

J. D. McMann's bay mare, Flora Temple,	1	1	1
J. L. Eoff's bay mare, Princess,	2	2	2
Time, first heat,			2.31
" second "			2.21
" third "			2.26½

Eclipse Course, L. I., June 6th.—Match $5,000; three-mile heats, to wagons.

D. Tallman's b. mare, Flora Temple,	1	1
J. L. Eoff's b. mare, Princess,	2	2

	FIRST HEAT.	SECOND HEAT.
Time of first mile,	2.37	2.37½
" second "	2.40½	2.36½
" third "	2.36½	2.45½

Eclipse Course, L. I., August 9.—Purse $1,000; mile heats; best 3 in 5, in harness.

  J. D. McMann's Flora Temple,   1   1   1
  J. L. Eoff's Princess,   2   2   2
    Time, 2.23½—2.22—2.23¼.

Nov. 21.—Proprietor's purse of $1,000; mile heats; best 3 in 5.

  J. D. McMann's Flora Temple, in harness, 1   1   1
  D. Mace's George M. Patchen, under saddle, 2   2   2
    Time, 2.28—2.23—2.24.

Nov. 24.—Purse $1,000; mile heats; best 3 in 5.

  J. D. McMann's Flora Temple,   2   1   1   1
  D. Pfifer's Ethan Allen,   1   2   2   2
    Time, 2.27—2.26½—2.27—2.29.

Kalamazoo, Michigan (on the National Park Course), *Oct.* 15.—Purse $2,000; mile heats; best 3 in 5, in harness.

  J. D. McMann's Flora Temple,   1   1   1
  J. L. Eoff's Princess,   2   2   2
  J. L. D. Eycleshimer's Honest Anse, 3   3   3
    Time, 2.23½—2.22½—2.19¾.

This is the best time on record, and probably never can be beat.

THE GREAT TROTTING MATCH.—*Flora Temple and George M. Patchen, the Emperor and Empress of the Trotting Turf.*— Few trotting matches could have created a like excitement and general interest. Here, on one hand, was the champion stallion, backed by a host of his friends, and here was the world-renowned queen of the trotting turf, with whose exploits the readers of her life and adventures, as penned by Mr. Wilkes, are familiar. Her friends had every confidence in her. They remembered her 2.19¾ at Kalamazoo, and backed her heavily at the current rate of odds, 100 to 80. Purse $1,000, at two-mile heats, in harness.

Just before the horses appeared, the odds that had been on Flora changed, and 100 to 80 was offered on Patchen. These offers were freely taken. The first appearance of Flora was the signal for loud cheering and applause, which Flora received, as usual, with an intelligent, quick nod of her head. Her eyes looked full of fire, and she was in most excellent condition, though somewhat heavy. Her driver, J. D. McMann, gave her two or three smart brushes up the home-stretch, to enliven her, and it had the desired effect. Patchen, who was driven by Darius Tallman, as in the two last matches, had the pole. The judges were Mr. Louis Brecal, Mr. Parks, and Mr. J. Hammil.

First Heat.—After scoring five or six times, owing to Patchen outfooting Flora, they got the word, and started even. immediately Flora broke, and Patchen opened a slight gap, but Flora, catching her trot, went quickly after him, and on rising the hill at the first turn she caught Patchen, and quickly changed positions with him. Flora passed the quarter-pole a length ahead, in 35 seconds, and going down the back-stretch she increased the gap, until near the half-mile (1.11), when Patchen again took the front, and before reaching the turn he had opened a space of a length between them. Flora here again came to her work, and with a pretty show of speed she glided upon him, until they trotted side by side. At this point the race was very exciting. The crowd surged to and fro, and a murmur of intense excitement and interest ran through all, which was increased to a wild and exulting cry, as the gallant little bay, when within two lengths of the judges' stand, encouraged by the lash and a loose rein, and the cheers of the anxious crowd, passed the stand a throat-latch in advance, in the extraordinary time of 2.21.

No sooner had the judges recorded the time on the board than a deafening shout of admiration and wonder arose from the immense mass of people that surged to and fro below the judges' stand. Their "little bay" had beaten even her own unapproached time on Long Island, and they had witnessed the most magnificent contest that had ever been seen on the Course. Betting was out of the question; every one seemed now to regard Flora's victory as a foregone conclusion. The betters who had backed Patchen at such long odds looked unutterable things, while the parties who had stuck to the Queen of the turf through good and evil report, were in raptures at her success. The usual time for cooling off elapsed, and the bell again signaled the horses for the

Second Heat.—After one false start, they got off even, and trotted steadily together to the turn. Here Flora began to quicken her step, and again was the Jersey stallion destined to see her flying heels. She led past the quarter-pole by an open length. Patchen followed earnestly after, but as if seeing the fruitlessness of his efforts, he took a fly into the air, and lost two lengths before he could be brought down. In going down the back-stretch he gathered together all his wondrous powers, and made after Flora as if determined to win or die in the attempt; but the pace was too much for him, and he again lost his footing, as Flora passed the half mile two open lengths ahead, in 1.11½. On the lower turn, the backers of Patchen had their hopes renewed for a moment by a plucky brush of Patchen, which closed the gap and placed him nearly side by side with Flora. Up the home-stretch they came nip and tuck, neither having the advantage, amidst the shouts and cheers of the multitude. They neither heeded friend nor foe, but swept on, with the goal in sight. They strained every nerve and power, and Flora won by a neck, in 2.24.

Third Heat.—Three false starts were made, and they were off. They remained neck and neck until the turn was made, when Flora took the lead, Patchen keeping close up. They passed the quarter-pole together in 36 seconds. In going down the back-stretch, Patchen closed and succeeded in passing Flora, and at the half-mile pole Patchen led; time 1.10. On the lower turn, Patchen was an open length in advance. Flora now commenced to make her work, and quicken her trot, as usual. She began to gain upon him. The contest was close and alternating until the distance-stand was reached, when Flora made one of her fearful efforts, which soon told in her favor, Patchen taking a fly in the air, when within two lengths of the judges' stand, and Flora was hailed the conqueror by the united voice of those assembled. The time of this heat was 2.21½.

And thus ended the great race. Flora has never done better, and when we consider that it was her first trot that season, this performance seems miraculous. In reviewing her career, we find the entire aggregate of the miles she won (249¼)—for we do not compute time for those in which she was not first—is 10 hours, 22 minutes, and 45 seconds; an average that could have been much reduced had she always been obliged to strive to win. The best average three single miles were those scored at Kalamazoo, on the 15th October, 1859, where she made her famous 2.19¾, in a third heat. The two previous heats were, 2.23½, 2.22½, thus making an aggregate of 7.05¾. Her next best average three single miles were those performed by her over the Eclipse Course, L. I., on the 9th of August, 1859, when she scored 2.23¼, 2.22, 2.23¼, making a total of 7 minutes 9 seconds for the three. The next best average was made in her three heats with Patchen, on the 21st November last, which add up 7.15. The Cincinnati trot, where she made a heat in 2.21½, adds up a second more; whilst the aggregate of this last trot foots up only 7.06½.

RULES TO BE OBSERVED IN CHOOSING A TROTTING-HORSE.

As the gift of speed depends upon mechanical principles, the trotting-horse may be chosen by measurement, and by observing these simple rules: The horse should be one fifth longer from the point of the shoulder to the stern than he is high; deep through the chest; a long and slanting shoulder, which enables him to elevate the arm, which should be long from the elbow to the knee, and short from the knee down. The back should be strong and rather short; rather narrow through the chest, and broad through the hips and stifle. In moving, the horse should carry his hind legs wide apart and his fore ones close together, and point them well forward. These simple rules never fail.

HEIGHT OF TROTTING-HORSES.

The annexed list gives the height of many celebrated horses, estimated only, but by two most experienced men, one of whom had groomed or ridden almost every one named, and the other is an old amateur, who had the quickest eye for a horse, and who rode after most of those named, and has seen them all repeatedly. Of the twenty-eight in the list, they differ only in regard to eight, and in those cases only by one inch, save in a single case. In these eight cases, we have given the estimate of the jockey who had ridden or driven them, and have great faith in its accuracy.

NAMES.	HANDS.	INCH.	NAMES.	HANDS.	INCH.
Dutchman	15	3½	Edwin Forrest	15	0
Lady Suffolk	15	2	Burster	15	0
Columbus	16	1	Norman Leslie	15	3
Aaron Burr	15	1	Confidence	15	2
Rattler	15	2	Locomotive	16	0
Screwdriver	16	0	Charlotte Temple	15	0
D. D. Tompkins	15	0	Washington	16	0
Lady Victory	15	2	Modesty	14	2
Topgallant	15	3	Greenwich Maid	15	0
Sir Peter	15	2	Awful	15	3
Whalebone	15	3	Paul Pry	16	0
Shakspeare	15	3	Ethan Allen	14	3
Betsy Baker	15	3	George M. Patchen	15	2
Cato	16	0	Flora Temple	14	1½

TRAINING THE TROTTING-HORSE.

The whole code consists in three words:—Air, Exercise, and Food; or, in other words, in bringing the horse under healthy conditions. The work given him in training should be severe according to his constitution, and consists in walking him from twelve to twenty miles daily, and giving him "sharp work" thee or four times a week. The sharp work is usually a distance of two and sometimes three miles. The horse should not be put to his speed this entire distance, but should be roused at intervals.

The horse should be brought gradually to his work, and his food increased with his exercise, which should be increased every day. The horse should never be pushed beyond his strength, nor ever driven off from his gait, so as to cause him to break or hitch in his gait. The whole secret of training the Horse consists in the gradual development of his muscles and wind, by judicious management as regards food, exercise, and grooming.

PATHOLOGY AND TREATMENT

OF THE

HORSE

AND OTHER

DOMESTIC ANIMALS.

INTRODUCTION.

In the prosecution of this work I have treated somewhat extensively on Pathology, or the nature and cause of disease, and the treatment of disease in general. The reader would do well to study these matters at his leisure, that he may understand treating disease on general principles. In treating of diseases under their separate heads, I have tried to simplify the subject so as to bring it within the comprehension of ordinary minds. I have given the symptoms and the treatment in the plainest and simplest manner, so that a prescription may be readily made out. I have given the designating symptoms only, in each disease, so that there can be no mistake in discriminating. As all diseases have many symptoms in common, were it not for this arrangement, we should be exceedingly puzzled in diagnosticating disease, or coming to correct conclusions as to the nature and cause of disease. I have embodied in this work the *modus operandi* of calomel, and the pernicious effects of minerals in general upon the animal system, with a treatise on bleeding, blistering, &c.

I have given credit in the body of this work, whenever practicable, to the authors from whom I have derived aid in the various departments of my labor; but I here gladly make an additional record of my indebtedness to the works of Morrell, Youatt, Fowler, and Beach, and to "The Country Gentleman," "The American Agriculturist," Wilkes' "Spirit of the Times," and the "American Trotting Journal and Racing Calendar."

GENERAL PATHOLOGY.

THERE are laws that govern in disease and death, as well as in life and health,—first principles in medicine as well as in philosophy, which, if understood, like the stars of the firmament guiding the mariner, will conduct the physician safely through the different stages of disease.

No branch of medicine is more interesting or useful than Pathology, considered either in a theoretical or practical point of view. By the term Pathology, we mean the history, nature, and cause of morbid action or disease.

There is no department of medical science involved in greater obscurity, less understood, or which has caused more controversy and investigation by different classes of physicians, in different ages, than Pathology. The conclusions to which they have arrived have generally been contradictory, unreasonable, and often absurd. Had the writers arrived at truth, by discovering correct principles, and resting upon simple facts, very different results would have attended their labors. For the want of this method of investigation, theory and speculation, unsupported by true physiological researches and observations, have been substituted for truth, and the inquirers have been led into error. I have collected from ancient facts, as well as from the light of modern science, all that was useful or practical; consequently, I am enabled to present a natural, simple, physiological, and practical system, based upon experience and observation.

Every man who knows much is learned; but he only is wise who has acquired practical knowledge, applicable in the common affairs of life. He endeavors to account for what he observes, and to discover principles in conformity with which he may constantly act.

It is very natural for us to conclude, at first view, that diseases are exceedingly numerous; but when we trace them to their elementary principles, we shall find that they are very few, or consist almost of a unit. The multiplicity of remote causes all so act upon the blood as to produce a very few proximate and direct causes, or a certain condition which we call disease.

All diseases resemble each other in their form, invasion, march, and decline. The type of all diseases is very similar as regards their pathological character. The different stages of disease have received (erroneously, however) different names, according to the different shades or symptoms assumed by the same disease. Physicians have heretofore been prescribing for symptoms, rather than for the original disease. Instead of laying the ax at the root of the evil, they have merely been cutting away at the branches. Their failure and want of success is the necesssary result.

In the course of future observations on Pathology, I shall divide morbid action or disease into two grand or principal causes; first, the remote or predisposing; second, the proximate or immediate

THE REMOTE CAUSE OF DISEASE.

Under this head I shall include all those great causes which act as predisposing agents, as heat, cold, and every other source of morbid influence or irritation. The atmosphere in which we are constantly immersed is full of danger. Both the organic and inorganic worlds of matter around us abound in poison. Sudden variations of temperature, excessive moisture or extreme dryness, different electric conditions, different degrees of pressure, a deficiency of light, or the atmosphere—which may be the source of disease, in consequence of being loaded with malarious contagion of various kinds, and noxious gases in general—may be considered as so many poisons.

IMMEDIATE OR PROXIMATE CAUSE OF DISEASE.

All the multifarious causes of disease mentioned as remote or exciting, produce a diseased condition of the blood, which is the proximate or immediate cause of disease. There can be no doubt that a corrupt state of the blood is the most frequent source of disease, by serving as a vehicle through which noxious substances are enabled to reach the parts upon which they act. Many poisons prove fatal by entering the circulation through the medium of absorption, and the miasmatic and contagious effluvia probably operate on the system through those channels. The blood, then, we see, is made up of many constituents, and is, of course, a compound, all the elements of which are in harmony with each other when in health, but often contaminated by foreign agencies introduced into it, and thus rendered unfit for the purpose of life; and, if it remains in the body, it passes from one state of degradation to another, and produces a great variety of disease.

UNEQUAL CIRCULATION OF THE BLOOD.

Having given the pathology of disease in relation to the vitiated condition of the blood, I will now speak of another prolific source of diseased action, which arises from a certain condition of it in relation to its circulation. The blood, in health, is equally distributed over every part of the body, conveying with it the elements of vitality and nutrition to every organ and tissue of the system; but when, from any cause, this natural and well-balanced state is disturbed, the equilibrium is lost, and the vessels from which it has receded become more or less collapsed, while the organ to which it has been driven becomes filled or congested, which prevents the due performance of its functions Consequently, this state creates irritation, inflammation, and perhaps ulceration. This effect is produced more especially when cold is applied to the surface of the body under certain circumstances, as in a state of perspiration, &c. It is known to physiologists that the whole amount of blood in the system circulates in a very short time through the extent of the capillary or small blood-vessels, ramifying near the surface, and that by those means deleterious agents are thrown out of the system, which, if retained, give rise to disease.

It is frequently during free perspiration, that cold comes in contact with the surface, causing these innumerable pores to close. One writer calculates these capillary tubes to be

twenty-eight miles in length, and believes that two thirds of everything taken into the stomach passes out by these emunctories. The effect, then, of these exhalents being suddenly closed or checked, when in a state of perspiration, is to impede their important functions, which prevents not only the perspirable matter from escaping, but drives back, with tremendous force, nearly the whole volume of blood from the capillaries to some one or more of the internal organs most predisposed to take on disease. The blood receding from the surface produces rigors or chills; the extremities become cold as well as the surface generally. Consequently, a large volume of blood is thrown in upon the internal organs, which, together with the irritation generated by morbid matter existing in the blood, causes pain and inflammation. When the blood is thrown upon the lungs and pleura, we have all the symptoms peculiar to inflammation of those organs. The condition of the system, then, points out with unerring certainty the course to pursue in the treatment of inflammatory diseases in general. And here I will pause for a moment to inquire what the orthodox system of practice is in those cases? It is the most irrational and absurd treatment that could possibly be adopted, consisting mostly in the abstraction of the *vital, irreparable balsam of life*—the blood.

FURTHER REMARKS ON THE NATURE AND TREATMENT OF DISEASE IN GENERAL.

I shall now inquire more particularly into the nature of disease, and lay down such rules as will lead to a successful mode of treatment. In prosecuting this inquiry no elaborate researches are deemed necessary; I have only to follow the precepts of reason and nature.

THE EXCRETORIES THE ONLY OUTLETS OF DISEASE.—The Author of our existence has wisely established certain laws in the animal economy, to guard and protect it against the inroads of disease, and, when formed, to remove it. Let us now inquire in what manner this is performed. A little attention to the system shows us that there are certain outlets or excretories—vessels designed to carry everything out of the system incompatible with health. When these excretories perform their office, health, or harmony of action, is the result; but when they cease to act, or act imperfectly, morbific matter is retained, and derangement or disease follows.

I shall briefly treat upon the several excretories, to show their office, and the consequences arising from their obstruction.

The Skin.—The whole body is covered and lined with this membrane, through which there are innumerable pores or openings, designed to carry off everything which is not salutary or compatible with a healthy state of the system. The fluid which thus passes off is distinguished into sensible and insensible perspiration. When the exhalents perform their office, and sweating or perspiration, sensible or insensible, is kept up and continued, the blood is pure, being separated in this manner from every impurity; but when the perspiration becomes checked by cold, the humors engendered in the system are retained, carried into the circulation, settled upon some organ that is most predisposed to take on disease, and become a source of irritation. Every day's observation convinces us that the moment the pores become in any degree closed, a universal derangement succeeds; a painful sensation is felt in that part where such retained perspiration is thrown, as if a needle or some foreign substance was piercing it. This may be said to be the proximate cause of disease. It may settle upon the lungs, the kidneys, brain, or some other organ, causing inflammation and pain; or it may be retained in the blood, and cause fever. This fact is demonstrated by the phenomena of eruptive diseases in general. The infection or contagion is taken into the blood through the medium of the lungs, and as soon as it becomes sufficiently impregnated with the specific humor or virus, nature is aroused and makes a powerful effort to expel it from the system. As soon as sho accomplishes this object, the poison in these eruptive complaints is thrown copiously to the surface, and appears in the form of vesicles or eruptions; and when they are thus expelled the fever subsides, but it will reappear, if from any cause the poison or humors are absorbed. These facts reduce it to a mathematical precision, and render the subject so simple and plain that it is really a matter of profound astonishment that any one the least acquainted with fever should be ignorant of its nature, cause, and cure, as well as those of other diseases. It is also well known that fever, inflammations, and a variety of other complaints immediately follow a checked perspiration.

The Bowels and Intestines.—The bowels or intestines are also designed by nature to carry off much that is injurious to the system, hence the diseases that arise from their constipation. It cannot be otherwise but such a great quantity of extraneous and feculated matter lodged in the body, and, perhaps, absorbents, must disorder it.

All parts of the intestinal canal are liable to the destructive agency of retained morbid excretions. These, if allowed to remain, not only act as a mechanical obstruction of a peculiar character but become the source of incalculable mischief. A disturbance of the functions of the whole frame must ultimately if not immediately be the consequence. We always find cathartic medicines possessing a most happy influence over many diseases of the most aggravated character, and find them necessary in almost every disordered state of the frame; they prevent the access of fresh symptoms, which almost invariably supervene when the equilibrium which seems to exist in the constitution is lost; and they prepare the system for the influence of other remedies, many of which, indeed, not only lose their power, but increase the mischief if those agents have not been duly and continuously given. Narcotics are only fresh sources of disturbance to the nervous system, tonics only act as astringents, and drugs whose powers are to be directed to particular organs lose their efficacy, if the bowels are not thoroughly cleansed.

The evils that are attendant upon an inattention to the due unloading of the bowels almost surpass the common belief. They are manifold, and almost every organ in the system sympathizes, directly or indirectly, with the digestive and excretive powers, and if they be disturbed sooner or later the ill consequences are manifested.

The Kidneys.—Through the medium of the kidneys the urine is secreted from the blood. This is another excretion designed to rid the system of foreign or morbific matter. When this excretion is checked, or if it does not duly perform its office, certain noxious matters are retained and mixed with the circulating fluids, and prove another source of morbid derangement, such as dropsy, inflammation, and how many other complaints it is difficult to describe. That diseases are carried off by a copious discharge of urine every physician knows. The effects which arise from the suppression of urine point out the purpose for which it is designed.

The Lungs.—Offensive agents are also secreted from the blood by the lungs; they not only throw off carbonic acid gas, but likewise mucus, and when they become diseased, more especially, they cast off pus, which, if retained, would cause suffocation. Hence we see in pulmonary disease an effort of nature to effect a cure through the medium of this organ.

Health, then, depends upon each and all of these performing their respective offices or functions. When any one becomes torpid, or ceases to perform its duties, morbid excitement, or disease is the consequence, and this shows most plainly the proximate cause of disease, at least, being nothing more or less than the retention in the system of morbific perspirable matter, producing irritation, morbid action, and a deviation from health. These humors are taken into the system through the medium of the air, food, or drink. The air breathed is returned loaded with watery vapor, which is calculated to amount to nearly forty ounces a day, from which we learn the injurious effects arising from its obstruction. We have a variety of symptoms in disease, but this is not owing to their exciting causes (these being similar), but to the peculiar structure or tissue which is the seat of the disease.

"The lungs," says Dr. Lamb, "are the great exhalents and ventilators of the blood; through them all morbid effluvia of the body are eliminated more copiously than by all excretory organs. The most virulent contagions pass out with the breath, and are diffused with it through the atmosphere." It cannot, therefore, be difficult to conceive that the pulmonary exhalation, habitually acrimonious or stimulating, should affect the lungs. The mucous membranes are excited to increased activity, in consequence of which the cough and expectoration become perpetual.

THE EFFORTS OF NATURE TO REMOVE DISEASE.

All men of eminence, all thinking men, all that are true to nature and themselves, both ancient and modern, depend much on the powers of nature to remove disease. It was called *vis medicatrix naturæ*, or a certain principle inherent in the system to expel from it every thing injurious, foreign, or extraneous. When a thorn is lodged in some part of the system, the first suggestion of the mind is to extract the thorn; but in the animal, where assistance cannot be had, or in man where it is beyond our reach, the design of nature is to bring on inflammation, and consequent suppuration, and thereby to remove the thorn. Should this effort be effectual, she next proceeds to the granulation of new flesh. The arteries and the veins, the lymphatics and the nerves, extend themselves, unite, and renew their communication, and, without the aid of a surgeon, nature effects a cure.

In pleuritic inflammation, nature pours forth coagulating lymph, and forms a new membrane, supplied, like the renovating flesh already mentioned, with arteries, lymphatics, nerves, and thereby preserves the substance of the lungs from injury.

Van Swimet makes mention of a case in which calculi in the gall-bladder being too large for the common duct, had, after producing inflammation, adhesion, and suppuration, found their way by fistulous ulcers to the external surface of the body, and thus effected their escape.

Yet more astonishing are her resources in case of necrosis; for, supposing some part of the bone to be deprived of animation, this she envelopes with new bone, united at each extremity with the fibers of the living bone. Here it proves a stimulant, and calls forth renewed efforts of the vital powers; inflammation is produced, suppuration follows, fistulous openings are formed in the new bones, and the dead portions of exfoliated bone are dissolved by the pus, and float off.

We see that the Author of nature has provided a principle which is calculated to remove disease. It is very observable in fevers. No sooner is noxious matter retained in the system than there is an increased action in the heart and arteries to eliminate the exciting cause by the exhalents or outlets of the system. With what propriety, then, can this provision of nature be denied, as it is by some? A noted professor in Philadelphia ridicules this power in the constitution; he says to his class, "Kick nature out of doors." It was this man, or a brother professor, who exclaimed, "Give me mercury in one hand and the lancet in the other, and I am prepared to cope with disease in every shape and form." I have no time to stop here and comment upon such palpably dangerous doctrine, but have only to say, Let the medical historian record this sentiment, maintained in the highest universities in America in the nineteenth century. Sometimes the energies of nature will cure disease in spite of their foolhardiness and misapplied drugs. They then put on their peacock-feathers, and ascribe to themselves that success which belongs to a strong constitution. Jann says, " As an idiot at the clock turns on, and only ends in stopping the machine, so rude practitioners continue working at the machinery of the Creator until, through their ignorance of the laws of nature, all further operations are suddenly suspended."

All the actions which we term *symptoms*, which are manifest during disease, are merely so many salutary processes set up by nature to remove some morbific agents which are present in the system, and consequently the great effort of the practitioner should be to aid these processes. We recognize it to be a principle or *law*, that medicine, to yield favorable results, must be given to act in harmony with the symptoms; that in a large majority of cases the efforts of nature are salutary—and that no reform in medicine will be effected until this is generally admitted; that practitioners, instead of resorting to a violent, acrid, and injurious medication, shall rather seek to wait upon nature, to pause where they do not see how they can assist her, and to offer aid upon all possible occasions. She knows what she is about, and can unite fractured bones, heal up wounded parts, call into action new organs when others are injured or destroyed, and, in short, perform so much, that for centuries the theory of a sentient principle superintending the functions of the body has always more or less been entertained.

The art of healing is very generally admitted to be one of those branches of medical knowledge in which there exists the greatest amount of errors, defects, and prejudices, and where experience is alike most difficult and deceptive. The mistakes that are daily made, are often far greater than we are willing to admit in the practice of our profession. It should ever be borne in mind that we have to do not only with the existing disease, but also with the conservative and reparatory efforts of nature, which by themselves often produce a cure. Hence the reputations of medicines and modes of treatment which so rapidly start up and are so quickly forgotten, and hence the false gods of therapeutics that to-day are adored and to-morrow are despised.

The one great principle, then, to which a comprehensive view of homeopathy, allopathy, hydropathy, and all other systems of medicine irresistibly leads, is, that in all cases and on all occasions, nature is truly the great agent in the cure of disease ; and as she acts in accordance with fixed and invariable laws, the aim of the physician ought always to be to facilitate her efforts by acting in harmony with and not in opposition to those laws. Disease is a mode of action of the living organism and not an entity apart from it. In accordance with this view, experience shows that when we make use of simple, natural means, recovery will ensue, in most cases, without the use of drugs at all. So far from being always necessary to cure, medicines are required only when the power of nature to resume her normal action proves inadequate, or is impeded by a removable obstruction. She may be *aided, but she ought never to be thwarted;* and medicine will advance toward the certainty of success only in proportion as we become saturated with this guiding principle.

We cannot treat disease successfully without studying its nature and cause. We must never forget that there is a mighty power always operating in our favor—this *vis medicatrix naturæ*. Do not thwart her beyond the mark, and she will get you through difficulties with which, without her aid, you could not cope. Nature, it is admitted, can repair injuries, restore lost parts, carry on a whole series of operations in the most delicate organs, such as the human mind can scarcely contemplate without becoming lost in wonder and admiration. But although the profession have long more or less recognized this wonderful internal sense, together with the fact that we have not only proofs that several actions which we call disease are instituted to get rid of causes producing them, yet the general ópinion seems to prevail that medical men are wiser than this internal sense, and that, although they daily see it carrying out operations, of the delicacy and beauty and absolute perfection of which it is hardly in the power of the human mind to form an estimate, yet most men will trust their families or animals in the hands of ignorant quacks, ignorant of the first principles of medicine, or the laws that govern the animal economy.

GENERAL INDICATIONS OF CURE.

If, then, disease is caused by retention of morbific agents, and a recession of blood from the surface,—if it be caused by morbid secretions,—if a deviation from a healthy standard is owing to the inactive state of the excretories,—do not the plainest dictates of reason and common sense show the necessity of restoring these secretions? Is there, or can there be, any other indication of cure, if they are the only channels which nature makes use of to restore the system to health?—and we confess we know of no other. The whole art of Physic, then, consists in aiding their salutary efforts. Then what else is there to accomplish, but to give such medicines as remove the obstruction, and restore the secretions. This is shown in the crises, or terminations of diseases in general. They subside when perspiration takes place, by diarrhœa, by vomiting, or urine, or expectoration. We ask, then, with what propriety the old school physicians give mercury, or bleed? Do they fulfill any indications of nature? Nature cures no disease by salivation, and seldom or never cures any by bleeding; hence, we may safely say that they are injurious. They produce diseases of a specific character. Bleeding lessens or destroys the healthy efforts established by nature, and thereby counteracts her intentions, and exasperates the complaint. An objector may say that nature often relieves herself by bleeding at the nose, hemorrhoidal vessels, &c., which must be admitted, but it is very rarely the case. But can it be proved that it is a healthy action established for this purpose? It is very evident that this is not the case. It appears rather to be the effect of disease—unequal circulation in the system,—and, therefore, instead of promoting this, as we should do if it was a natural excretion, we find it good practice to check it by equalizing the circulation. Bleeding is not a choice of nature, but a forced condition, which is the only alternative. In all cases, if proper evacuations of the excretions had been made in time, nature would not have had to resort to hemorrhage. Scarcely any diseases are terminated by a discharge of blood from the system; but they almost invariably subside by purging, sweating, and urine. Then, if health is restored in this manner, is it philosophy—is it common sense—is it acting in the capacity of a servant of nature, to institute a mode of treatment which she seldom or never takes to accomplish this object?

THE PHYSICIAN CAN ONLY BE THE HANDMAID OF NATURE IN THE CURE OF DISEASE.

We can cure nothing in reality; we can only remove the offending cause, and assist nature to perform a cure; and, therefore, lay it down as a fundamental maxim in medicine, that all the physician can do, is to act as a servant of nature.

The ancients not only observed the effects of the instinct by which brutes are directed to certain plants for relief, when they are unwell, and then applied them to the complaints of man; but they also attended with diligence to the manner in which nature, when left to herself, relieves or throws off disease. They perceived that certain disorders were carried off by spontaneous vomiting; others by looseness; and others by augmented perspiration; and having thus learned how diseases were cured by nature, whenever her powers seemed too weak and tardy, then, and only then, they ventured to assist her by art, by accurately observing all the motions, endeavors, and indications of nature to carry off disease, and by observing by what critical evacuations she casts off morbid matter which causes them, and so restores health. We may by the same method of reasoning know both the methods and the means we should use to assist in producing those salutary effects.

By thus investigating, observing, and truly knowing the diseases and their causes, we shall know when to assist nature according to her indications; and in this consists the chief part of medical knowledge. And when we shall thus have learned of nature, by observing her laws and indications, we may reasonably hope to render the theory and practice of medicine beneficial to the world.

MERCURY.

History.—It is not generally known that mineral poisons, or mercurial preparations, in any form, have been in use as a medicine only three or four hundred years. Such, however, is the fact. For five thousand four hundred and ninety-seven years these deadly agents were unknown to mankind, and well would it have been for the world if they had forever remained unknown. During all this long period the Empiric and Galenic systems alone were resorted to as curative agents—being chiefly vegetable or botanic. From these remedical agents, which had been generally potent in the removal of disease, the attention of the world was suddenly called in the year 1493, to that of mineral poison as a curative agent. About that time Theophrastus Bombastus Paracelsus first taught in Switzerland that mineral poison might and ought to be used as a medicine, and all that administered it were denominated "*quacks*," in allusion to the name "quicksilver," given this metal by the Germans. This individual succeeded in overthrowing the *Galenic* System, which had stood the test for fourteen hundred years, and in its place he introduced the Mineral or Chemical System. The introduction of mineral agents into medical practice caused great excitement. The regular physicians of that day, the Galenic or Botanic, contended with zeal against minerals; while on the other hand, the chemical practitioners or quacks inveighed against botanics as being weak and inefficient. The whole medical world was thus kept in commotion for two hundred years; both sides assailed each other in the most opprobrious epithets, and the contest has continued to the present day. Since the days of Paracelsus the great mass of physicians have placed their chief reliance upon the lancet, the knife, and a few acrid and poisonous minerals.

Mercury is met with in the metallic state in the quicksilver mines of South America. It is principally brought from Almaden, in Spain; from Idria, in Illyria; and from Moschel, in Bavaria; and there are extensive mines in California, the most noted of which is worked by an English Company.

Mercury, in its pure state, is fluid; and, from this circumstance, together with its likeness to silver, it has obtained different names expressive of these characteristics; hence the English word Quicksilver, or living-silver, and the Greek name *Hydrargyrus*, or water-silver. It is the most brilliant of metals, and one of the most weighty; and such is its expansive power, that if it be confined in an iron globe, hooped solidly together with prodigious strength, and then heated in a furnace, it will burst its inclosure into fragments, and escape.

A most remarkable instance of the effects of quicksilver is related in an extract from a letter from Lisbon, dated May 12th, 1810, in the sixth volume of the "Edinburgh Medical and Surgical Journal." In April, 1810, the "Triumph," man-of-war, and the "Phipps," schooner, saved from the wreck of a Spanish ship, off Cadiz, a large quantity of quicksilver. The

"Triumph" took on board thirty tons, contained in leathern bags of fifty pounds each. Those bags were packed upon shore. Saturated with sea-water, they were collected and stowed away in the bread-room, after-hold, and store-rooms forward. In about a fortnight many of them decayed and burst, and the mercury escaped into the recess of the ship. At this period, bilge-water had collected; the stench was considerable, and the carpenter, in the act of sounding the well, was nearly suffocated. The common effect of the collection of bilge-water is to change, from the escape of the gas, every metallic substance in the ship to a black color; but on this occasion every metal was coated with quicksilver. An alarming illness broke out among the crew, all of whom were more or less salivated. The surgeons, pursers, and three petty officers, who were nearest the place where it was stowed, felt the effects the most; their heads and tongues having swollen to an alarming degree. The ship was sent to Gibraltar to be cleared, and the people were placed in the hospital. The quantity on board the "Phipps" was not so great. She was sent to Lisbon to be cleared, by boring a hole in her bottom and letting the quicksilver run out. Mouse and cockroach on board were under Allopathic treatment, and the symptoms of general salivation appeared in a strong degree. Of the truth of the statement, Dr. Bird and the surgeons bear sufficient testimony.

Of the effects upon miners, we may likewise judge by a narrative that is given by Dr. John Wilkins in the "Philosophical Transactions," London, in the year 1666, in which he described the quicksilver mines at Treuli, in the Venetian territory. He says that although none of the miners stay under ground more than six hours in a day, all of them die hectic, or become paralytic. He saw there a man who had been in the mines only six months before he became so filled with mercury that on putting a piece of brass in his mouth, or rubbing it between his fingers, it immediately became white like silver, and precisely in the same state as if mercury had been rubbed upon it; and so paralytic was the unfortunate man that he could not, with both his hands, carry a glass to his mouth. It appears that both the shaking palsy and salivation are the consequences of exposure to the vapor of the metal in its usual state, but that those that are liable to the one are not so to the other; the same exposure in one causing salivation, and in another palsy. Of this,

Dr. Christion furnishes us with an illustration, which he learned from a friend, Mr. Harding, the mineralogist. A barometer-maker and one of his men being exposed one night to the vapors of mercury from a pot on the stove, in which a fire had been accidentally kindled, they were both most severely affected with salivation, which caused the loss of their teeth, and a shaking palsy which lasted all through their lives.

Dr. Falconer, of Bath, England, gives us an account of the effects produced by the application of this metal in the form of a girdle about the waist, especially by females of the lowest order, for the cure of the itch, as being a cleanlier proceeding, and more free from fetor than ointments composed of sulphur. Many cases were admitted into the Bath Hospital, and the symptoms which were exhibited were a degree of general weakness approaching to palsy, great pain and tremor in the limbs, and often violent headaches. It is worth remarking that an instance lately occurred in the Bath Hospital, where all the symptoms that distinguish the poison of lead were observed, even the loss of tone in the muscles of the wrist, in consequence of the use of mercurial ointment for the cure of the itch.

Dr. Bateman, in his history of the disease to which mirror-silverers are subject, says: "The attack is sometimes sudden; at others, gradual. It begins with unsteadiness and shaking of the limbs, which prevents walking, speaking, or mastication. The tremors become frequent; every action is performed by starts; if the occupations that produced it be continued, sleeplessness and loss of memory supervene, and death terminates the scene. A peculiar brownish hue of the whole body and skin generally accompanies the disease. In its first attack it may be taken for St. Vitus' dance; in its latter stages, for delirium tremens."

General Opinion.—So penetrating and deleterious are the effects of mercury on every part of the system that there is not a tissue but it reaches and injures. It passes into the most minute vessels and there remains, proving a source of irritation. It not only affects the soft parts, including the nervous system, but enters the bones, which it decomposes, and causes exfoliation.

The most remarkable effect of mercury is its action on the salivary glands—*salivation*. When this medicine is introduced into the system in such a manner as to excite this peculiar state, at first it produces increased vascular action, shortly followed by a metallic or brassy taste in the mouth, and a slight mercurial fetor of the breath ; the gums become swollen and spongy, and their edges soon present a slight degree of ulceration ; the lining membrane of the cheeks and palate acquires a leaden hue and is swollen, and an increased flow of saliva takes place, accompanied by pain in the teeth on the least pressure. If these symptoms be allowed to advance, by giving more mercury, the cheeks, the tongue, and the throat swell and ulcerate, and a copious flow of saliva, sometimes amounting to several pints in twenty-four hours, is induced. This excessive salivation is accompanied by slow fever and rapid emaciation.

The effects of mercury on the system are sometimes accompanied by a peculiarly alarming state, first described by Mr. Pearson, of England, under the name of *mercurial erethismus*. It is characterized by great depression of strength ; a sense of anxiety about the præcordia ; frequent trembling ; a small, quick, and sometimes intermitting pulse ; occasional vomiting ; a pale, contracted countenance ; a sense of coldness ; but the tongue is seldom furred. When these symptoms are present, a sudden and violent exertion of animal powers, such as rising suddenly, will often prove fatal. The use of this mineral is also frequently attended with or followed by several forms of disease of the skin Of this the most important is *mercurial eczema*, which often occurs when only a small quantity of mercury has been taken. In its milder form it resembles the acute stage of *eczema rubrum* arising from other causes ; but it more frequently assumes a much more severe character, when it is ushered in by a fever, difficult respiration, dry cough, and tightness across the chest, with a general smarting and burning sensation of the skin over the whole body. These symptoms are soon followed by an eruption of minute vesicles, which break and discharge a very fetid fluid. As the disease increases in severity, the eruption extends over the face and whole body, which become covered with incrustations ; the fever assumes a typhoid type ; the difficulty of breathing increases, and is accompanied by bloody expectoration ; spots of purple appear ; and death ensues, preceded by delirium and convulsions.

The evils resulting from the employment of mercury, says Dr. J. King, of Kentucky, are not confined to the observations of a few, but are reiterated again and again by nearly all medical writers, or at least all that have become eminent in their profession, and indisputably establish the fact that this mineral ought never to be used as a medicinal agent. Nevertheless it is still resorted to by the physicians for almost all diseases; nor can we be surprised at it, when we are presented with such a medley of theories, and such discordant practice, as the various professors of medicine issue from time to time. But do physicians understand how or why mercury produces its action upon the system? We have conversed with many physicians upon this subject, and could never obtain from them any more satisfactory theory than that its irritating quality, or its peculiar mode of action, is the cause. This is certainly an incomprehensible reason for him who honestly wishes to aim at the truth; and the very next question would be, How is this peculiar mode of action produced? or how, acting merely as an irritant, so many serious effects should result?

If we refer to standard authority in this matter, we still remain in obscurity. Thus, in the "United States Dispensatory," under the article Mercury, it is said : " Of the *modus operandi* of mercury we *know nothing*, except that it probably acts through the medium of the circulation, and that it possesses a peculiar alterative power over the vital functions, which enables it in many cases to subvert diseased action by substituting its own in their stead."

If such, then, is all the knowledge which physicians have of the action of this mineral—such all the satisfaction we can derive from them—and yet, notwithstanding this ignorance, they still continue to employ it, I ask, Is this not downright empiricism? or, if not, what is? In Eberlie's Therapeutics, we find the following remarks:—

" Mercury, it is observed, by Cullin, acts as a stimulant to every sensible and moving fiber of the body. What the peculiar character of the excitement which it produces may be, *it would be in vain to inquire*, but it appears to be more permanent and universal than that of any other medicinal agent with which we are acquainted.

" We will now demonstrate the peculiar character of the excitement which mercury produces on the system, as ascertained by physiological and chemical results, notwithstanding the bugbear assertion, it would be in vain to inquire.

" 1. Phosphoric acid pervades almost every fluid and solid of the living body, and is more abundant than any acid.

" 2. Phosphoric acid and lime, in the form of phosphate of lime, constitutes the greatest part of the composition of human bone.

" 3. Phosphoric acid does not act upon mercury, but combines with its oxyd, forming phosphate of mercury.

" 4. Phosphorous acid differs from phosphoric acid in containing one proportion less of oxygen, and decomposes all the oxyds and salts of mercury, separating the mercury into its metallic state.

" 5. Bile, which is secreted by the liver, is composed of water, albumen, pricromel, muriate of soda, phosphate of soda, phosphate of lime, soda and lime uncombined with an acid, &c. Hence the chemical character of the bile is alkaline.

" 6. Saliva is composed of water, mucus, animal matter, alkaline muriates, lactate of soda, and pure soda.

" 7. The mucous secretion, from the mouth throughout the whole alimentary canal, with the exception of the gastric and pancreatic juices, as has been proven by M. Downe and others, is of an alkaline character.

" 8. It is admitted by all chemists that acids and alkalies mutually decompose or neutralize each other, forming new combinations; also, that the affinity existing between acids and alkalies appears to be much greater than that between any other known substances in nature.

" 9. In whatever soluble form the usual preparations of mercury are introduced into the stomach, they are reduced to an acid of the metal previous to being absorbed into the system.

" 10. Most acids are capable of combining with the oxyds of mercury.

" By keeping the above facts in view, we can clearly understand the *modus operandi* of mercury.

" Thus, when any salts of mercury—say a dose of calomel, for instance—have been taken into the stomach, as soon as it has passed through the lower orifice of the stomach (pylorus) into the first intestine, or, as sometimes and more properly termed, the second stomach (duodenum), it comes in contact with the bile, which is discharged from the liver into the bowels at this point. Here, in consequence of the affinity existing between the acid combined with the metal and the alkali of the bile, the acid is separated and forms with the alkali a new combination, possessing three elements, while the mercury is left in the form of the black oxyd, which is the natural oxyd of this metal. And it may be opportune to mention here that any salts of mercury, when exposed to the action of the atmosphere for a sufficient length of time, will result in the black oxyd.

"By the above action, a mutual decomposition takes place, both of the sub-muriate of mercury and the phosphate of lime of the bile, upon the principle called, in chemical language, electuary affinity. Thus, with the soda and the lime have a stronger affinity or chemical attraction for the muriatic acid of the calomel than the affinity existing between the muriatic acid and the mercury. Hence the soda and the lime combine with the muriatic acid, forming muriate of soda and lime, while a portion of the phosphoric acid is set free by a decomposition of the phosphate of lime, and the mercury, by losing its acid, is reduced to an oxyd. This is not only proved by chemical laws, but also confirmed by physiological facts; and in this manner calomel works or acts upon the liver. We may likewise understand why the discharges from the bowels produced by mercurial cathartics are invariably dark colored, like the black or gray oxyd of mercury. In the form of an oxyd, then, is mercury carried into the mass of blood, to be thence circulated to every part of the system. Combining with the phosphoric acid of the bones, a phosphate of mercury is formed, leaving the bone in the state of an oxyd of calcium or common lime; the bony structure being thus chemically decomposed, crumbles and exfoliates.

"A similar combination with the phosphoric acid of the nerves and brain produces nervousness, severe pains, loss of memory, headache, &c.; and as the changes of the atmosphere act upon mercury in any state, the suffering patient can predict the various changes about to take place in the weather with as much precision as could be derived from the most delicate barometer.

But pure mercury, or mercury in its metalic state, has been found in various parts of the bodies of those who have used it as a medicine, by several celebrated anatomists; and how could this have been produced from its phosphate? I reply, Not from the phosphate of mercury, which salt, whenever found, separates after a time, freeing the phosphorous acid and leaving the mercury in its metallic state.

"Phosphorous acid may be produced by any element capable of abstracting from phosphoric acid a part of its oxygen. Phosphorous acid being thus formed and coming in contact with the oxyd of mercury, will form a phosphate of mercury, from which eventually the mercury will be precipitated, or separated into its metallic state, in which state it may remain an indefinite period of time. This also affords a clue to an understanding of the statement made by Dr. Goldsmith, in his Natural History, that those miners who have been condemned to labor for life in the mercurial ores, often transpire quicksilver at every pore before death releases them from their sufferings.

"The oxyd of mercury is capable of producing decomposition, to some extent, in every fluid or solid in the human body. And if any gentleman of the old school can disprove the foregoing explanation of the *modus operandi* of mercury, I trust you will allow him the use of the pages of your journal (Western Medical Reformer, Cincinnati), that is, if he dares to risk his reputation or expose his ignorance by attempting it."

"Among the numerous poisons," says Dr. Hamilton, "which have been used for the cure or alleviation of disease, there are few which possess more active, and, of course, more dangerous powers than mercury. Even the simplest and mildest forms of that mineral exert a most extensive influence over the human frame, and many of its chemical preparations are so destructive, that in the smallest doses they speedily destroy life."

Practitioners of the first standing, on every trifling occasion, prescribe calomel or the blue pill. Thus calomel is now almost the universal opening medicine recommended for infants and children; and a course of blue pill is advised without any discimination for the trifling irregularities of digestion in grown persons. Dr. Falconer, of Bath, has, in strong language, reprobated this practice, and has pointed out many of the dangers of the use of mercury. His warning voice, however, has not been listened to, for the employment of mercury has for several years become more and more extensive. When the effects of mercury upon the human body are accurately investigated, and duly considered, it will cease to be given as a medicine.

It is the object of the author to illustrate, in the following pages, these propositions; and in doing so, he readily avails himself of the recorded facts and observations of the distinguished members of the profession. In detailing the changes produced upon the system by preparations of mercury, it is necessary to premise the well-known facts that there are some individuals on whom such medicines, though contained for considerable time, have little or no perceptible influence, unless the activity of their form, or the magnitude of the dose be calculated to excite immediate effects. For example, when the constitution of the person is feeble and very susceptible, a very few grains of the muriate of mercury, given in substance, prove rapidly fatal, and large doses of the submuriate are quickly followed by vomiting and purging. On the other hand, instances of constitutions which are unsusceptible to the influence of the ordinary doses and preparations of mercury, are very few in comparison with those which are affected by the smallest quantity of that mineral. Preparations of mercury exhibited internally for any length of time increase in general the action of the heart and arteries, and produce salivation, followed by emaciation and debility, with an extremely irritable state of the whole system.

The effects of mercury are expressly mentioned or virtually admitted by every author, ancient and modern, who has directed its use, and it must appear very extraordinary that their full influence should have been misunderstood, or at least not sufficiently regarded.

Blood drawn from the arm of the most delicate individual subjected to a course of mercury, exhibits the same buffy crust as blood drawn from a person laboring under pleurisy, and the secretions from the kidneys or excretions from the skin are greatly increased.

From the time that the influence of mercury becomes evident, the general strength declines rapidly. It appears, therefore, that the increased action of the heart and arteries, excited by mercurial medicine, produces not only the same injurious changes upon the body with those arising from inflammation; but also certain effects peculiar to itself. This important fact has been incidentally noticed by numerous authors, although the natural inference to be deduced from it has been very much overlooked. Dr. Carmichal expressly says, " Mercury produces a specific fever, different from all others, and attended with an increase of the various secretions." Salivation, or an excessive and unnatural flow of saliva, in general, follows the increased action of the heart and arteries; producing a certain metallic taste in the mouth, and attended with a peculiar odor from the breath, different from what is ever perceived in a natural disease. The excessive flow of saliva in consequence of mercury is accompanied with more or less local inflammation of all parts within the mouth; in some cases, besides the ordinary ulceration of the gums and loosing and final separation of the teeth, the tongue, palate, &c., swell and ulcerate to a frightful degree. Emaciation so commonly follows a course of mercury, that several eminent physicians, about the beginning of the last century, imagined that mercury had a natural tendency to destroy the fatty particles. The celebrated Van Swiston says, "All the pinguid humors are disolved by the action of mercury, all viscid are attenuated and discharged out of the body together with the virus adhering to them; therefore, the patient's body is totally emaciated." Dr. Blackwell has proved that the serum of the blood passes off with the urine; it is more than probable that the excessive rapidity of the emaciation is occasioned by that circumstance. Debility, with an irritable state of the whole system, accompanies the emaciation, and of course occurs in various degrees in different individuals. The late Benjamin Bell, whose practical knowledge was so pre-eminent, comprehends in one small paragraph an emphatic list of those effects of mercury. He remarks that, besides the usual symptoms of fever, "mercury excites restlessness, anxiety, general debility and a very distressful state of the whole system."

The consequences of this effect upon the nerves are different upon different subjects. In some, temporary delirium takes place; in others, palsy or epilepsy supervenes; and in many, the memory and judgment are more or less impaired. Instances have occured where sudden death has supervened in consequence of a very trifling exertion or agitation. Delicate individuals, however, particularly those that have been accustomed to sedentary life, and therefore, in a special degree, females, generally experience after a course of mercury various modifications of disordered feeling which unfit them for the duties of life, and render existence a burden.

Among the complaints arising from this cause may be enumerated impaired appetite, with all the ordinary symptoms of indigestion, particularly retching and flatulency in the morning, depressed sleep with frightful dreams, impaired vision, frequent aches and pains in different parts of the body, occasionally such sudden failure of strength as if just dying, and at other times violent palpitations of the heart, accompanied with difficulty of breathing; along with all those complaints there is a wretchedness of look, with such a propensity to brood over their miserable feelings, that it renders life a curse rather than a blessing.

"I might cite all the Materia Medica," says Dr. Falconer, "for authorities that the long-continued and frequent use of mercury is not free from danger; that among other ill effects it tends to produce tremors and paralysis, and not unfrequently mania. I have, myself, seen repeatedly, from this cause, a kind of proximation to those maladies, that embittered life to such a shocking degree, by the nervous agitation with which it was accompanied, as to make it probable that many of the suicides which disgrace our country were occasioned by the intolerable feelings that result from such a state of the nervous system." To the truth of these remarks every unprejudiced physician who has been in extensive practice must bear testimony.

Another consequence of the use of mercury is a very violent eruption of the skin. This eruption is usually preceded by heat and itching, a frequent pulse, and a white tongue. Most commonly it begins on the inside of the thighs, or about the flexure of the arms; and Dr. Pearson asserts that it generally attacks the anterior parts of the body before the posterior. The parts affected are first of a faint red color, and gradually the shade becomes deeper. The eruption proceeds by slow degrees over the whole surface, accompanied with an evident tumefaction of the skin, with great tenderness and heat, and a most troublesome itching.

The minute vesicles contain at first a pellucid fluid, and are each surrounded by a circular redness. From the great itching they are soon ruptured, and discharge a thin acrid fluid, which irritates and excoriates the surface, and aggravates greatly the patient's sufferings. In this way the disease proceeds from one part to another till the whole person becomes affected. Dr. Ally has described its various forms by different names, such as hydrargyria, mitis, simplex febrilis, and maligna; and out of forty-three cases which he witnessed within ten years, eighteen died. And I have become fully convinced that nine tenths of all the cutaneous or skin diseases are caused by this and other mineral poisons used as medicines, and all the most formidable diseases that I meet with are produced by improper medication. Dr. Falconer thus mentions that he once saw mercurial remedies used for a redness in the face, which they effectually removed, but instantly produced a dropsy of the chest, terminating in death. Dr. Blackall has recorded similar cases. Dr. Ally asserts that he has seen an eruption appear over the entire body of a boy about seven years old, for whom but three grains of calomel had been prescribed ineffectually as a purgative. Many other instances of the violent effects from a small dose of mercury might be cited. Besides, the following seems to prove that mercury may remain inert for a considerable time in the system, and afterward, by some unknown cause, may become active.

A lady in the twenty-eighth year of her age had a bad attack of hemorrhage. She was very much reduced from the loss of blood. Three days after, she complained of a bad taste in her mouth, with soreness of her gums, and on the following day salivation took place. All she had taken was wine and peruvian bark. On inquiring into her history, it was learned that four years before she had for a fortnight a course of the blue pill, which had only slightly touched the gums, and she asserted that she had never again taken any preparation of mercury, and had in general been in good health. The salivation, with the usual consequences of excessive emaciation and debility, continued for twelve months.

It is universally acknowledged that although the morbid effects of mercury may be induced very suddenly, and by small quantities of the medicine in certain constitutions, there are no marks by which such peculiarities of habit can be distinguished, and there is no method of arresting their progress. Hence the great danger of using this destructive mineral. Had these injurious effects of calomel upon the delicate constitution been hid from the rest of the profession, and known only to the author, some apology might be offered for the pertinacity with which that medicine is still prescribed; but so far from this being true, it may be confidently asserted that no medical man of competent knowledge and observation could administer calomel in many instances without being convinced of its injurious tendency. Of this, numerous proofs could be cited, but it is sufficient to appeal to the testimony of Professor Carlisle and Dr. Blackall. Mr. Carlisle has expressed himself very strongly on this subject: "That grave men should violently persist in large doses of calomel, and order their doses to be daily reiterated in chronic debilitated cases, is passing strange. Men starting into the exercise of the medical profession from a cloistered study of books, and from abstract speculations—men wholly unaware of the fallibility of medicinal evidences, and unversed in the doubtful effects of medicines—may be themselves deluded, and delude others for a time; but when experience has proved their error, it would be magnanimous, and yet no more than just, to renounce both the opinion and the practice." .Dr. Blackall's remarks, being very specific, afford a still more satisfactory proof of the validity of the author's opinions. "It appears to me," he says, "that no accident proper to the disease can account for all those fatal conversions to the head, which of late years have so frequently taken place in the fevers of children; and I have on some occasions been disposed to attribute them to excessive and repeated doses of calomel, which, not moving the bowels as we expected, have either given evidence of being absorbed, or, on the other hand, have purged too violently, and been succeeded by diarrhea without bile, and a prostration of strength, from which the little patient has never risen.

Its less effects are sometimes of no slight importance—a slow and imperfect recovery, a languid feverish habit, and a disposition to scrofula.

When we reflect that in fevers mercury is given with little scruple, we are led to remark that within the last thirty years either a sudden revolution in the laws of the human machine has taken place, or that medical men have ceased to reason on the operation of medicine.

Every practitioner who has paid the least attention to the effects of mercury in fevers, must admit its immediate and subsequent injurious effects. No further proof need be adduced than is found in the preceding pages. It is shown that it produces an augmentation of feverish symptoms; that from the time that the influence of the mercury becomes evident the general strength declines rapidly, and a dangerous emaciation and debility takes place, with an irritable state of the whole system; and paralysis, epilepsy, loss of the senses, and many other distressing and dangerous complaints follow.

Who, then, in the possession of his reason would think of exhibiting mercury in fevers? But strange as it may appear, it is universally administered, and constitutes the chief medicine in the materia medica. We hope, however, that those who see its pernicious and fatal effects portrayed in this work, will in future fly from it as from the bite of the most poisonous serpent. It would require years, and volumes ten times larger than the present, to give an account of the number of deaths it has occasioned.

It has the power of decomposing the bones and causing exfoliation. Dr. Mathias states that mercurial disease is much worse than that awful scourge and curse, the venereal disease. "I have seen," says he, "the cartilages and bones of the nose and palate all confounded in one diseased mass. I have also seen several cases of the mercurial disease, in which the complaint first begins in the nose, and after having produced considerable destruction there, the ulcerative process has crept upon each side of the jaw-bones, through the cheeks, in an irregular direction, till at last the miserable patients have found the remedy for their sufferings only in death. This disease is never known to produce this effect unless treated by mercury, as has been lately proven in France by the French physicians."

"This is an era of calomel," says Dr. Anthony Hun, of Kentucky; "the present medical practice might well dispense with every drug besides it. I own that the calomel practice is both cheap and easy to physicians, for the whole extent of both theory and practice is, Give calomel; if this will not help, give more calomel; and if that again proves abortive, double, treble, the dose. If the patient recovers, calomel has cured him; if he dies, nothing on earth could have saved him." The reader will conclude that medical schools and academics, with the head-aching-studies of Anatomy, Physiology, Botany, Pharmacy, Chemistry, &c.; have been laid prostrate by this giant, Calomel. Half a day's—nay, in a *genius*, half an hour's study will initiate any lady or gentleman into all the mysteries of the Esculapian art, and the *aurea praxis* might swell the account of a modern Galenus to one dollar at the expense of twelve and a half cents. This is certainly for the doctor a consummation devoutly to be wished; but there is a heavy drawback on our joy, which the fable of the boys and the frogs so ingeniously portrays. What is joy to you is death to us, said the expiring frogs. Dr. Sweetler on Consumption, states that calomel has often been ranked among the causes of consumption; he says that it acts as an exciting cause of tubercles. Mercury saps the constitution; creates the very disease which it is given to cure; and lays the foundation for infirmity, suffering, and premature decay. If mothers or doctors deal out calomel to their children or patients, we can only recommend them to the mercy of Heaven. Dr. Beach, of New York City, was called to visit a child to whom a physician had administered mercury, and says that such a horid melancholy spectacle he never witnessed. Nearly all one side of the face, eye, and neck, was mortified black, and destroyed, by this mineral; and the wretched child was then dying from its effects. An attempt was afterwards made to get the doctor indicted, but without effect, as the patient was killed according to rule and custom. Suppose this mischief, or rather manslaughter, had been commited by a reformed practicioner; how soon would he have been arrested, committed, and punished! While ignorance continues, this evil will continue; but just as soon as the community is enlightened on the subject, down goes the poisoning system. What can change the tyranny of custom and fashion, or the religion of the Turks or the Chinese? Nothing but intelligence and the light of Heaven.

Are there any of my readers who would not by this time pray, "Deliver us from Calomel?" You may, fellow-citizens,— you can—drive away prejudice, that black thunder-cloud which ever hovers over truth. Think for yourselves; consult your precious health and lives. Every man should, at least to a certain degree, be his own lawyer, his own preacher, his own doctor. He should not hire the preacher to do up his religious thinking,—the doctor his physiological thinking. For what is man when he makes himself a cowering, cringing slave to the opinions of others, and tamely bows down to win the momentary smiles of popular applause! He who goes through life leaning entirely upon books and the childish fancies of others, without thinking for himself, renders his present life a blank, inasmuch as he lays his head in the dust without its having bequeathed one original thought to the world, for the benefit of after generations.

4

My method of curing disease is entirely without mercury and its doleful effects. Were I not more successful than those gentlemen of the medical profession, who trust so much to the virtues of calomel, still the gain would be immense; but, from facts enumerated fairly by myself and others, I have nothing to fear from an impartial comparison.

Now I have given you the *modus operandi* of mercury, or its effects on the human system; and, as the Horse exists by similar laws, and is subject to similar diseases, the effects will be similar on this animal. Dr. Beach, in his book on the Theory and Practice of Medicine, has an engraving of a dog's stomach and intestines, irritated and inflamed in consequence of taking calomel. Although the stomach of the Horse is less sensitive than man's, in consequence of which there is no medicine that will cause vomiting in the Horse, yet the lower part of the intestinal canal is far more sensitive than in the human subject. In consequence of the extreme sensitiveness of this portion of the intestines we have to be very careful in administering drastic cathartics, and especially calomel, which will invariably produce irritation, and, if carried too far, inflammation in the bowels.

MINERALS GENERALLY.

Having exhibited in the preceding section the pernicious effects of that Sampson of the materia medica, Mercury, which has slain its hundreds of thousands, I shall now merely hint at the dangerous effects of some of the other minerals used for medicine.

Zinc.—Zinc is the next mineral extolled by writers as a suitable article for medicine. The following definition may be given of it: Zinc is a metal which exerts a powerful and very dangerous effect upon the system, especially if an over-dose is taken.

Antimony.—This metal is given the most extensively of all minerals which the mineral doctors have in use. It is known that Antimony is given to vomit; in all inflammatory diseases to relax and to produce a moisture on the surface; and this is the article applied to the skin, mixed with lard, to produce pustules or eruptions, and after applying it a few days they appear and cause the most poisonous and painful ulcers.

Now it is obvious that the same mineral, when given internally, must act as a poison, by irritating the stomach and bowels;

and is there not danger of its causing pustules in these organs as well as on the surface? I believe, too, that cancers in the stomach and intestines are produced by this and other mineral poisons administered as medicines. Again, antimony being very soluble in water is liable to be absorbed in the circulation, and exert its destructive irritating and poisonous effects on every organ, causing a metallic taste in the mouth, nausea, vomiting, hiccough, burning heat and pain in the stomach, colic, copious evacuations from the bowels, fainting, increased action of the heart and arteries, cold surface, difficult respiration, loss of sense, convulsions and death. And notwithstanding all these baneful effects, physicians are in the habit of administering this dangerous metal. Antimony, says Hooper, is a medicine of the greatest power of any known substance. A quantity too minute to be sensible in the most delicate balance, is capable of producing the most violent effects if taken dissolved or in a soluble state.

Arsenic or *Ratsbane*—This is another mineral which physicians of the old school are in the habit of giving as a medicine, while it is known that a few grains are sufficient to destroy life. It is usually disguised and given in the form of Fowler's Solution, which is very pleasant to take. It is also applied externally in the form of powder or plaster for the destruction of corns, and in this way is sometimes absorbed, and proves fatal. Given internally it produces nausea, sickening, burning near the heart and over the whole body, inflammation and eruptions on the face, lips, tongue, palate and throat, vomiting, black and filthy stools, small pulse, palpitation, great thirst, fainting, coldness, cold sweats, difficulty of breathing, bloody urine, swelling and aching of the body, livid spots on the surface, a great prostration, loss of sight, delirium, convulsions, and sometimes it proves fatal. It has been shown by dissection that the stomach and bowels have by it been inflamed and ulcerated, and partly destroyed. It is known that this mineral is given to destroy rats, and yet persons take it because physicians prescribe it. It is stated by Hooper that arsenic is one of the most sudden and violent poisons we are acquainted with. When the quantity is so small as not to prove fatal, tremors, palsy, and lingering hectics succeed. We have the combined testimony of many medical practitioners conspicuous for their professional zeal and integrity; and I am irresistibly induced to declare my opinion at least, against the internal use of this active and dangerous medicine or poison.

Metals Generally.—The attention of the reader is plainly directed to the following testimony of the above medical writer: All the metallic preparations are uncertain, as it depends entirely on the state of the stomach whether they have no action at all or operate with dangerous violence.

"Minerals," says the learned Dr. Choyne, "are the most destructive agents to animal bodies that malice can invent, beyond gunpowder itself and spirituous liquors; for nature has provided none such but as poisons in venomous creatures to kill their enemies. They become iron bristles, nails, and lancets, darting into their solids, so as to quickly tear, rend, and destroy, and can never, therefore, be proper for food or medicine; whereas Galenical or vegetable productions have none of those bad properties, and are therefore better designed for food and medicines."

BLOOD-LETTING.

Among the various means made use of to restore the sick to health, there is none so irrational and absurd as blood-letting. It is considered at present as a universal remedy, and resorted to for the cure of the slightest indisposition, and is daily slaying its thousands; still it continues to be the main pillar of the profession. Indeed, were bleeding and mercury to be altogether prohibited, many physicians would find themselves in a sad dilemma, for their hands would be tied. We are unable to determine precisely the commencement of this pernicious custom, but we find it to be very ancient, having been, it appears, cotemporary with the declension of the healing art in the earliest ages of the world. Sydenham, it appears, gave a new impulse to this practice about three centuries ago, and is supposed to have introduced it as a remedial agent; but this is contradicted by medical history. It was not carried, however, to such an extent till after the discovery of the circulation of the blood by Harvey. It was at this time that the whole faculty began their mad career in committing the most wanton violation of the laws of nature. Those who are so unfortunate as to fall victims to disease are doomed to suffer the most extravagant effusion of blood, and the poor sufferers are soon hurried to an untimely grave, *secundem artem.*

Even the guillotine of France scarcely surpassed this systematic murdering. But in process of time practitioners began to witness the mischief they were committing, which in some measure damped their ardor in these bloody scenes. This check induced one physician to remark, that the proportionate disuse of the lancet was one of the greatest improvements in modern medicine. We find, however, that blood-letting has been practiced for many centuries, almost with the same infatuation; and, lamentably for mankind, in the present day it is regarded as a most powerful weapon to subdue disease. There are few maladies in which it is not recommended. In pleurisy, and in all inflammatory complaints, an astonishing quantity of blood is drawn from the system. It is very common to take from five to seven pounds in twenty-four hours. One of the professors in the medical college in the City of New York stated that he had frequently bled his patients to the amount of two hundred ounces in three days; another professor declared that he had taken three hundred ounces in a short space of time, and for proof of this fact appealed to one of his students. The effect of this practice I shall leave for people of common sense to determine. How much is it to be regretted that such an awful scourge of humanity should exist! A little examination into the consequences of blood-letting will prove that, so far from its being beneficial, it is productive of the most serious and fatal effects. Nature has endowed the animal frame with the power of preparing from proper aliment a certain quantity of blood. This vital fluid, subservient to nature, is, by the amazing structure of the heart and blood-vessels, circulated through the different parts of the system. A certain natural balance between what is taken in and what is passed off by the several outlets of the body, is, in a state of health, regularly preserved. Blood-letting tends artificially to destroy that natural balance in the constitution. Nature deprived of a quantity of this circulating fluid, leaving less means for repairing the loss she has sustained, begins immediately to repair it. The secretions and excretions in general are diminished, the appetite is increased, and for a short time the process of nutrition is unusually quick.

Thus, by the wisdom of Providence, nature restores to the constitution what art had taken from it. The consequences, therefore, of having been once bled are rarely considerable.

This simple operation is, however, an imprudent violation of nature and common sense.

But too often the practice has not rested here; for various are the incidents which favor the reputation of blood-letting. The patient, if addicted to an easy, indolent, luxurious way of living, may find himself after the evacuation sensible of present ease. The system, being before too full of blood, or rather, the balance being lost, enjoys a short respite from its usual oppression. Or after the bleeding, though it was improper and tended to increase the disease, yet, the hope of relief, or a change of weather, the benefit of exercise and country air, or some other alteration in an accustomed manner of living, may, by palliating or removing the complaint, prejudice the patient in favor of the lancet. The disorder, it may be, was of such a kind as really to admit of alleviation from the use of bleeding; but, nevertheless, the remedy unhappily proves of worse effect to the constitution than the disease itself would have done, though entirely left to nature. Great numbers of people, who have been relieved by bleeding, are apt to be partial to the supposed means of their own recovery, and to become strenuous advocates of its use, even in cases by no means similar to their own. These and a variety of other accidental causes often persuade to repetitions of blood-letting. The consequence now become more serious. The constitution, though it did not suffer materially from one bleeding, yet, far from being able to undergo with impunity repeated operations of the same kind, turns against itself those powers which were given for its preservation, and co-operates with the imprudent use of the lancet in promoting the accomplishment of its own destruction. For now the constitution not only repairs the losses of blood it sustains, but, if the common intervals of time be interposed, makes more blood than is naturally required for the purposes of health and life, that it may be able to bear such repeated evacuations. Thus the habit of blood-letting is established.

But in fact habitual blood-letting augments the very evil it was intended to remove; for sanguine evacuations necessitating the constitution to make more blood than is requisite, produce too great fullness of the system. The balance between what is taken into the body and what passes off by its several outlets is no longer maintained. As the disposition to plethora itself, if the person continue to live in his accusomed manner, will undoubtedly prevail, except at that time when the constitution has just recovered the unnatural assistance of the lancet, the habit of blood-letting increases, and becomes stronger by repetition. In this state, the constitution, in spite of human art, will at times labor under various degrees of plethora, till the vessels arrive at that point of fullness which again creates the necessity of bleeding. Though some constitutions are so robust or so peculiarly formed by nature as to bear such treatment without any immediate bad consequences, yet these are but the privileged few. Many will severely suffer, though they themselves may often be the first to extol, in the highest terms of praise, that very remedy which has proved so pernicious to their own constitution. They have been bled till stated bleedings become necessary, not only for the support of health, but even for the preservation of their lives. They have injudiciously created the habit of bleeding, and are even happy to find that it gives present relief in complaints which from the first it tended to induce, and afterwards to confirm.

The effects of plethora are many and dangerous. A slight degree of it often produces strange commotions in those of weak and irritable habits. No person who depends for the preservation of health on an artificial discharge of blood, can ever be pronounced out of danger. Before the usual means of relief can be employed, the sanguine fullness, at one time or another, may be produced to a morbid degree, or even lead to a fatal termination. The anticipation of the stated bleeding may, with the greatest inconvenience, lessen, but it can never remove danger. An increase of secretions, unnatural heat, torpor, inactivity, and a sense of lassitude, are common affections of plethora. The whole vascular system is unnaturally put upon the stretch, and along with it the nervous and muscular fibers. Thus by slow degrees the tone of the body, in consequence of so considerable an over-distention, is in danger of being destroyed. The constitution itself, in proportion to its natural vigor, is rendered liable, exclusive of every other cause of disease, to break many years sooner than it otherwise would in the common course of nature, if nature's laws had not been wantonly violated, or presumptuously despised. Hence, old age sets in at an earlier season, accompanied with heavier infirmities. Frequently the appetite fails, the powers of digestion and nutrition are impaired, the body shrinks, the mind becomes dejected, the stomach and bowels become disordered, sleep is interrupted and unrefreshing, and, in short, the whole constitution fundamentally shaken and debilitated.

These are the slow but frequent consequences of bloodletting. Others indeed occur which, though on the whole, perhaps, less destructive, are, however, more painful and better distinguished.

The following are the observations of the surgeon of the western regiments of Kentish militia in England on the effects of blood-letting; they plainly prove the inutility and absurdity of bleeding for the "cure of any disease with which we are acquainted." His language ought to be written in letters of gold. He remarks: "I have been upwards of six years a surgeon of the western regiment of Kentish militia, during which time our number of sick has never been inconsiderable, whereby much opportunity of practice has been afforded me. I have been in the habit of keeping a journal of the different cases as they occurred, wherein I carefully noted every symptom of which the patient complained; the various remedies exhibited; the time when, and with what view given. I also marked every change that took place in the course of a disease, and the effects of medicine made use of, and lastly my own opinion of the method of cure which I adopted. In the course of my practice I have endeavored on every occasion to determine the justness of preconceived theories by experience, and on every subject to think for myself, uninfluenced by tenets of schools or the opinions of others. The prevalence of any mode of practice is certainly not a clear proof of its being useful, nor is it a sufficient recommendation that it may be practiced with safety. If it is not evidently beneficial it ought to be laid aside. In this light I condemn the custom of bleeding as a means of cure in febrile and other diseases, which I have no hesitation in asserting is not necessary in any complaint. If we grant that any deviation from the healthy state denotes debility, either general or partial, surely whatever has a tendency to debilitate further, it is reasonable to suppose, ought to be carefully avoided. It certainly cannot be denied that in every disease where bleeding has been practiced, complete recovery has been protracted owing to the debility thereby occasioned. We are directed to use blood-letting to lessen irritability, to tear off the *phlogistic diathesis*, to replenish blood vessels, and to prevent inflamma-

.on. I know by experience that those indications can be fulfilled much better, with much less danger, by other means. Though the ill effects of the loss of blood, unless excessive, are seldom perceptible in youth, yet they rarely fail of being felt before the age of forty. People who have been often bled when young, about this period of life begin to be affected with chronic pains, they recover very slowly from fits of illness, and are very liable to febrile paroxysms and a variety of other disorders. I have rarely been deceived in my conjectures respecting patients of this description, when I have met with them. The cases mentioned by Dr. Denman show that it does not prevent inflammation or abortion, nor is it proved that by taking away blood we lessen the diameter of the blood-vessels, for we find that six ounces from a large orifice have a greater effect than twenty-five from a small one."

Dr. John Pully remarks: "There are evil symptoms following the use of the lancet, not depending on the action of any morbid poison, not resting on the unscientific conduct of the operation, but owing their appearance to a peculiarity (call it irritability if you please) of constitution. Sometimes an abscess forms in a cellular membrane around the puncture from the lancet, which commonly appears the size of a walnut, and if the habit be very bad the inflammation will extend far around and a considerable sloughing of the parts may be the consequence, insomuch as to render the removal of the limb a matter of necessity; and even after amputation the stumps will in all probability assume the like disposition to slough. In either case the symptoms of irritation may be great enough to destroy life. When the vein is disposed to inflammation, much pain is felt after bleeding, and around the punctured part shortly appears a redness and swelling which soon extend along the arm both above and below the elbow. The arm feels knotty, and pain is given by touch. The inflammation and swelling sometimes extend to the breast. The accompanying symptoms of irritation are always great, sometimes producing delirium, and even the death of the patient." Horses, after bleeding, are frequently attacked with this affection of the brain, which the author has witnesssed on several occasions. Did physicians know the number of people killed by bleeding, I am persuaded they would abolish such an irrational practice. It always endangers the life of the patient, and never fails to aggravate his disorder; and if so fortunate as

4*

to recover, he experiences a train of evil circumstances through life.

The blood is properly called the vital fluid, and the life of a person is said to be in the blood. We know that just in proportion to the loss of this fluid, by accident or the lancet, is our strength taken from the system; it is succeeded by great prostration of strength, and a derangement of all the functions of the body. These effects are invariably, in a greater or less degree, consequent on bleeding. It is not unreasonable to suppose that what will debilitate the strongest constitution in a state of health, will be attended with most serious evils when applied to a person laboring under any malady. " But," says one, " we take away the strength to arrest the disease; in other words, we must make the patient worse before we can make him better."

This argument shows how ignorant medical men are of the animal economy, and the indications and cure of disease. No system could be invented better calculated to counteract the healthy efforts of nature. Bleeding is immediately resorted to in all inflammatory complaints; but did practitioners know the nature and design of inflammation the treatment would be different. Fever is produced by an increased action of the heart and arteries, to expel acrid and noxious humors, and should be promoted until the irritating matter is expelled from the system. Fever is a wholesome and salutary effort of nature to throw off disease or some morbific matter, and, therefore, every means to lessen this indication proves injurious. Bleeding, in consequence of the debility it produces, prevents such indications from being fulfilled. If the employment of the lancet was abolished altogether, it would perhaps save annually a greater number of lives than in one year the sword has ever destroyed. Medical men are sometimes apt to consider themselves, and are generally regarded by others, as inefficient, unless they are doing something—that is, either performing some painful operation, or administering some powerful remedy; whereas the fact is, that in a considerable proportion of cases, the best thing that can be done is to let the patient alone.

If the reader wishes to learn anything more of the effects of blood-letting in disease, I would refer him to the high authority of Dr. Marshall Hall, of England, who has written a book on the morbid and curative effects of blood-letting.

I have now given you something of a knowledge of the effects of blood-letting, in health and disease, on the human subject; and the Horse exists by the same law, and is nourished by the same principle, which is the blood. The effects of blood-letting are the same in all animate beings, and the life is taken in the same proportion as we take their blood. The arterial blood, or vital fluid, is the only source from which the whole body derives its nourishment and support, which is taken from the nutritive properties of the food. Thus we see the absurdity of depriving the system of any of this vital principle.

FEVERS IN GENERAL.

The word *fever* is derived from the Latin term *febris*, which signifies to be hot, or to burn; it is applied to a class of diseases characterized by morbid heat of the skin, unnatural frequency of the pulse, and disturbance in the various functions. Fever is a natural and salutary process, indispensably necessary to throw off a poison generated in the body, or induced by external causes, irritation, or obstructions. There is an increased action of the heart and arteries, to expel from the system this irritating or morbific matter, or to bring about a healthy action. It is often fatal; but this is rather to be attributed to the fault of the constitution than to the disease, or rather to improper treatment.

Foredice thus describes fever: "A general disease, which affects the whole system—the head and the trunk of the body, and the extremities; the circulating, absorbing, and nervous system; the skin, the muscular fibers, and the membranes; the body, and likewise the mind."

Fever commences by shivering or rigors, followed by hot skin, a quick pulse, and a feeling of languor or lassitude, loss of appetite, thirst, restlessness, and diminished secretions. These constitute the leading symptoms of fever, the characteristic features by which its presence may always be detected. Every function of the body is more or less disturbed. Fever constitutes perhaps the largest proportion or class of disease which assails the human family or animal kingdom. Notwithstanding the numerous inquiries, experiments, and theories on the subject by medical men, from time immemorial, the treatment remains the same among the old-school physicians. But there is at this day no uniformity in either opinion or practice among the different classes of medical men. They all go blindly to work to cure it like the physician mentioned by Dr. Alombert. He compares him to a blind man armed with a club, who comes to interfere between nature and the disease. If he strikes the disease, all is well; if he strikes nature, he kills the patient. Says a writer :—" This is the disease which to break, to baffle, to conquer or subdue, the learned College of Physicians have tried all their efforts and spent their skill in vain. It must run its course, is the common sentiment. If one mode of treatment fails, we must try another, and another, till the exhausted imagination, the worn-out resources of the *materia medica*, and the dying patient arrest the hand of the experimenter (and I might have said the tormentor), or nature triumphs equally over medicine and disease. The very medicine which aggravates and protracts the malady binds a laurel on the professor's brow. When at last the sick are saved by the living powers of nature struggling against death and the physician, he receives all the credit of a miraculous cure; he is lauded to the skies for delivering the sick from a detail of the most deadly symptoms of misery into which he had plunged them, and out of which they never would have risen but by the restorative efforts of that living power which at once triumphed over poison, disease, and death.

The causes which have conspired to cover with uncertainty the treatment of fevers, and to arm the members of the faculty often against each other, are numerous and important. A brief detail would unfold the many causes of error, and the fatal consequences which often result from the established practice.

Dr. Eberly, author of a treatise on the theory and practice of medicine, thus remarks:—" When, indeed, it is considered that the destroying angel has made his most desolating visitations under the form of febrile epidemics, and that in the long list of human maladies *fever* occurs in perhaps nine cases out of ten, the paramount importance of this subject is strongly forced upon our convictions." " If we except," says Van Swient, "those who perish by a violent death, and such as are extinguished by means of old age (and which are indeed few), almost all the rest die either of fever or of diseases accompanied with fever.

We read in Pliny with what fear and trembling the Romans endeavored to have this universal disease (fever) appeased by their supplications in the Temple of Fanum. And hence, perhaps, it is that fevers are called disease by Herod, and that Horace called all diseases simply fevers, when they rush out of Pandora's box."

Dr. Donalson, who published a new Theory and Practice of Fevers, remarks as follows:—" From a retrospective glance at the history of our science, we are forced to acknowledge that there is perhaps no subject which is more eminently calculated to humble the pride of human reason than this one; for in relation to this subject, especially, pathology has been in a continued state of revolution and instability. The human mind has been engaged with this subject for nearly three thousand years. Theories have risen and fallen again in continued and rapid succession. Each has had its hour to strut upon the stage, and its votaries to yield it faith; but the stream of time has hitherto overturned all these unsubstantial though highly-wrought fabrics.

In fact, no physician of the old school whose works I have read—no professor of medicine whom I have heard speak on the nature of disease—has ever discovered, or even hinted, at the nature and cause of fevers; all have delivered theories which amounted to open acknowledgment of their ignorance of it, or have candidly confessed their universal ignorance respecting the nature of disease.

Fever may be considered a *friendly* effort of *nature* to expel morbific agents, by the skin, or other outlets or excretories, from the system, or to contend against the action of external agents or predisposing causes. Symptoms, which we call disease, ought to be treated as friends, and not as enemies. Thus, all the varied grades, forms, and developments in which disease presents itself, are nothing more or less than a continued warfare between nature and offending causes; and when she has gained the victory over her enemy a healthy action is established, an equilibrium takes place, and nature again enjoys a season of rest, and again goes on peaceably superintending her own domestic affairs. This fact is demonstrated by the phenomena of eruptive disease—small pox, measles, &c. The infection or contagion is taken into the blood, through the medium of the lungs, and as soon as it becomes sufficiently impregnated with the specific humor or virus, a preternatural action of the blood-vessels immediately takes place. Nature is aroused, annd makes a powerful effort or struggle to expel the poison from the system. As soon as she accomplishes this object, the exciting cause or agent in those eruptive complaints is thrown to the surface and appears in the form of vesicles or eruptions; and when they are thus expelled, the fever immediately subsides, but will reappear, if from debility or other causes the poison or humor is re-absorbed. All contagious diseases resemble each other, as regards their pathological character, whether in man or animal. In the Horse we have a class of diseases termed by farriers Horse Distemper, Pink-eye, Tongue-ail, Epidemic Catarrh, &c. All develop themselves differently, according to the specific nature of the poison, or the gland or organ on which it locates or takes effect. Thus we see disease presents itself in a variety of forms. The different grades or stages of disease have received (erroneously, however) distinct names, according to the different shades or symptoms assumed by the same disease.

But, says the objector, one disease may run into another; for instance, a bilious *fever* into a typhus or a congestive *fever*, &c.; but when a doctor attempts to play that game on you, just tell him that he is using deception, or exposing his ignorance. Fever is fever until it is cured, or until nature triumphs over her enemy, and a healthy action of the secretions and excretions is established; then disease can no longer exist, fevers will not continue to set in one after another, until the patient has had all the fevers that their imaginations can invent.

Writers on Farriery have described a variety of *fevers* in the Horse; but their observations appear to have been drawn from the works of medical authors on the human subject, and their reasoning seems entirely analogical. I can distinguish only three kinds of fevers in the Horse: the first an *idiopathic*, or original disease, and therefore properly termed *simple;* the second dependent on internal inflammation, and very justly denominated *symptomatic;* the third caused by atmospheric influence falling on a large number at once, and consequently termed *epidemic* fever. For example, if the lungs, bowels, or stomach were inflamed, the whole system would be thrown into disorder, and a *symptomatic* fever produced; but if a collapse of the perspiring vessels happens to take place, the blood will accumulate in the internal parts of the body, and though inflammation is not produced by it, the unequal distribution of the blood alone will occasion that derangement of the system which constitutes *simple* fever. Again, if the animal is brought under certain epidemic influences, as soon as it becomes sufficiently impregnated with the specific poison or virus to cause an irritation, nature makes a powerful effort to expel the poison from the system, and produces a variety of symptoms which we call disease, and this I call *epidemic* fever.

TREATMENT OF FEVERS IN GENERAL.

COMMON TREATMENT.

The principal and almost only remedies, if such they may be called, resorted to by physicians of the Allopathic school, are mercury, antimony, blood-letting, and blistering. Instead of rendering nature the necessary assistance, her powers and energies are entirely crushed, weakened, or diminished, thereby corrupting the fluids and inducing another dangerous disease worse than the original. I am satisfied that their mode of treatment in febrile, as well as all other diseases, brings on a contaminated state of the blood, and dangerous, if not fatal debility. This practice is certainly absurd and irrational; and I ask, Is it not preposterous in the extreme? It cannot be supported by argument, reason, experience, or facts. I shall not consume time here to descant on the impropriety and the injurious effects of such practice. These have been explained under their proper heads.

REFORMED PRACTICE.

General Indications of Cure.—Restore the suppressed evacuations or the secretions and excretions. This will remove the offending cause; and when this is removed, the effect, or, in other words, the fever, will necessarily cease. In fulfilling these indications consists the whole secret of curing disease.

Practical Indications of Cure.—First, Equalize the circulation. Second, Moderate the violence of the arterial excitement. Third, Obviate local inflammation and congestion. Fourth, Support the powers of the system. Fifth, Relieve urgent symptoms. The necessity of fulfilling all these indications must be borne in mind by the practitioner. In every modification of fever it becomes his duty to render himself an assistant of nature. What she endeavors in the commencement of the disease to accomplish is to evacuate the deleterious agents by the proper passages. The whole business of art, therefore, is to assist in these two efforts of secretion and excretion of morbific matter.

The manner of effecting this, in every particular disease, is given under its respective head; but as we are treating of fevers in general, it may be proper to give the principles of treating them without entering into their various modifications. The treatment here laid down is proper for all grades of inflammatory disease, whether local or general. The remedies which are to be given to assist the secretions and expel morbific matter, are, sudorific diluent drinks, to produce a determination to the surface; cathartics, to evacuate and give a healthy tone to the stomach and bowels; diuretics, to operate upon the kidneys, &c. As the skin, bowels, kidneys, and lungs are the only avenues or outlets of disease, a cure can be effected only on the principle of stimulating these excretories to action. For this purpose we give sudorifics to produce perspiration; diuretics to stimulate the secretion of urine; cathartics for the bowels; expectorants for the lungs; stimulants to increase the action when too low; sedatives to allay irritation when too high. These are the only medicines or classes of medicine that are necessary in all forms or types of disease. Physicians have classified, divided, subdivided, and enumerated diseases almost beyond the power of computation; and however wide they have differed in the names of diseases, all have to depend on those six classes of medicines—stimulants, sedatives, diaphoretics, diuretics, cathartics, and expectorants, in the cure of all forms of disease. Consequently the whole art of treating disease consists in the judicious selection and proper application of these six remedies. In this, and this alone, consists the whole art of healing.

When we are called to see a patient, the first object in view should be to ascertain causes, and remove them if practicable and the effect will cease. Unless causes can be removed, our treatment can only be palliative. Seek to ascertain the first departure from nature's laws, or the first organ that ceases to perform its function, and direct your medicine to that organ first. If the kidneys be defective in their operation, give diuretics; if the bowels be disordered, give cathartics; and if the skin or the capillary system does not perform its office properly, give sudorific or sweating medicines, &c.

There is no specific for fever. We should make no favorite prescription our hobby, and give it without regard to conditions; if we do, we shall fail in our attempt to cure disease. When medicine cures disease, it is always given under favorable conditions or circumstances.

The sanative or healing principle is in the animal, not in the medicine. The following will illustrate my meaning on this point:—A fruit-tree being heavily laden with fruit, in consequence one of its branches is partially severed from the trunk. The gardener enters, props up the limb, binds it up with adhesive bandages, while the tree is left to the inherent operaations of its own sanative principle. This principle being in the tree, it must heal itself. This operation is under the control of its vegetive life. So with a fractured bone, after the splints and bandages have been applied, this operation is under the control of the animal life. The healing of an ulcer, the curing of disease, and all the healing operations in animal or vegetable life, are effected on this principle. Medicines only serve as props or bandages, while nature does the cure.

The Bowels.—The intimate relation which exists between the whole of the alimentary tube or canal, the skin, and other parts of the animal economy, points out the necessity of promoting in them a healthy action. Purgatives, therefore, have a decided good effect in fevers. The preternatural excitement of the blood-vessels is sensibly diminished by the exhibition of purgatives; this effect takes place by removing the feculent matter which they contain, and by stimulating the exhalent vessels of the mucous membrane of the intestines, causing them to pour out copious effusions from the blood or circulating mass.

The Pores of the Skin.—If the pores of the skin, or the capillary system, has received a check from taking cold, the perspiration being thus obstructed, has to be reabsorbed and thrown back on some internal organ, causing irritation and derangement. This is the cause of a large majority of all fevers. It appears that febrile diseases in their very nature consist in a derangement of the skin or capillary system, and no manner of treatment will cure a fever until this function is restored. It is calculated that two-thirds of all which is taken into the system, is evaporated by sensible and insensible perspiration; hence, it will be seen what mischief will arise from a retention of this fluid and what benefit will follow by restoring this secretion.

Another great benefit to be derived from perspiration, is the evaporation that continually takes place on the surface, which keeps the skin cool and soft.

The Kidneys.—When the kidneys cease to perform their office, or do it imperfectly, the urine is scanty or much diminished; this fluid is retained and carried into the circulation; and hence the necessity of restoring their secretion. Diuretic medicines, therefore, or such as promote a discharge of urine, must be administered; and it will be found that, as the urine begins to flow freely, there is diminished arterial excitement. It often is the case that fevers and other inflammatory diseases are brought to a favorable termination by a spontaneous discharge of this fluid.

MODE OF PROMOTING PERSPIRATION.

I have already hinted at some of the means of promoting perspiration, but it may be proper to dwell somewhat more upon them in this place. In general, perspiration may be promoted by taking warm diluent drinks, such as catnip, sage, balm, or saxifrage, freely given; if these fail, give freely of the sudorific tincture; or, if the extremities are cold, attended with shivering, composition tea as warm as can be borne; and in extreme cases, bring the patient under the influence of *veratrum viride*,—dose 30 drops of the fluid extract. (These medicines, and the mode of administering them, will be explained under their proper heads.) At the same time, bathe the feet in buckets of warm water. Any bucket can be used, by applying an additional bottom cut out of a board and put under the real bottom, the thickness of the chime; this arrangement will prevent the horse's weight from pressing out the bottom. A leather bucket is more convenient; this may be made of any thick leather. The bucket should be made six inches in diameter, and long enough to come up to the knees and hock-joint, and there secured by means of a strap and buckle. India rubber is a more convenient material for the bucket; those that intend following farriery would do well to keep those convenient articles on hand.

On the application of the foot-bath the horse must be clothed with blankets extending to the floor; then set a kettle of hot water under the horse, and keep up the steam by throwing hot bricks into the kettle of water. After sufficient sweating wrap the horse closely in blankets all over, remove the buckets, rub the legs until dry, and wrap them in flannel.

The application of the foot-bath is the safest, surest, and most effectual way of equalizing the circulation, and should be applied in all cases when reaction cannot be got up by friction and counter-irritants. The foot-bath is one of the most valuable applications that can be made use of in the diseases of animals. I can cure more diseases by the simple operation of the foot-bath, and the means of promoting perspiration, as above described, than the old-school doctors can with their whole materia medica.

SIMPLE OR PRIMARY FEVER.

There should be a distinction made and kept in view between primary and symptomatic fever. First, primary, when it does not arise from any other complaint; second, when it does arise from some other complaint, such as injuries, wounds, &c. Some have denied that simple fever exists in the Horse, but they must have been extremely careless observers of the diseases of that animal.

The simple fever does not occur so frequently as the symptomatic, nor is it by any means so formidable in its appearance; yet it is necessary to give it the earliest attention, for unless nature receives timely assistance, she will be sometimes unable to get rid of the load which oppresses her; and the blood will accumulate in the internal parts of the body, until inflammation in some of the viscera is produced, and a dangerous disease is established.

Symptoms.—Shivering, loss of appetite, dejected appearance, quick pulse, hot mouth, and some degree of debility; the horse is generally costive, and voids his urine with difficulty. The disease is often accompanied with quickness of breathing, and in a few cases with pain in the bowels, or symptoms of colic.

Treatment.—A check of the perspiration being the cause of the disease, the first object should be to restore this excretion to its natural state. To effect this, the Sudorific Tincture No. 1 is the most effectual. Give one ounce of the Tincture in a pint of warm water or thin gruel once in six hours, until perspiration is established and the circulation becomes equalized. If this fails of success, recourse may be had to the foot-bath, and manage as directed under the head *Mode of Producing Perspiration.* After this object is accomplished, if costiveness be one of the symptoms, give the Cathartic Powder No. 2 in a pint of gruel; a clyster of warm castile soapsuds may be injected, and the physic and injections repeated once in six hours until you get a free operation from the bowels. After the operation of the physic give the Diaphoretic Powder No. 9, morning and evening, until the fever abates and the horse appears more lively. For want of the above medicines, a strong infusion of catnip, balm, sage, or saxifrage may be given; or some other sudorific medicines, as composition-tea, made by pouring a quart of boiling water on one ounce of the Composition Powder No. 7, to be given as a drench at one dose. Warm water, bran mashes, sodden oats, are to be frequently offered; warm clothing, to keep up insensible perspiration, frequent hand-rubbing, and a liberal allowance of litter, are all necessary. After the fever has subsided, if the horse's appetite does not return, give

Powder of golden seal, . . ½ ounce,
Ginger, ... 1 "

in his feed once a day.

The tinctures and powders can be found under the proper heads. For general principles of treating disease, look under the head of *Treatment of Fevers in General.* I have been more particular in describing this class of diseases, for there is scarcely a disease to which flesh is heir that will not appear upon a strict examination either to consist in or to be a consequence of inflammation, which, when it attacks any of the internal organs, gives rise to the most dangerous diseases. Thus, an inflammation of the lungs, bowels, or any of the internal parts, will produce that kind of derangement in the system which is termed fever. As fever is a necessary consequence of inflammation, the author has not given it a separate consideration; if the heat is confined to a single organ it is called inflammation; when general, fever; which will be treated of more extensively under the head *Symptomatic Fever.* In the long list of maladies, fever occurs in perhaps nine cases out of ten. Consequently, the paramount importance of this subject is strongly forced upon our consideration. I have given the subject of fevers a thorough investigation; my researches have been long and studious, and I have labored hard to know all that could be learned, both from ancient facts and the light of modern science, on the human as well as the animal subjects. I have also given you the views of some of the most eminent of the medical faculty, together with the conclusions and practical experience of the author, which, if understood, will conduct the practitioner safely through the different stages of disease.

SYMPTOMATIC FEVER.

Symptomatic Fever is one of the most common diseases to which the Horse is liable. In a state of nature, or when living under natural conditions, the Horse is the most free from disease of all animals; but when domesticated, and subjected to our use and abuses, he is the most liable of all animals. Nor need we be surprised, when we see the amount of labor imposed upon him and the extraordinary feats he is compelled to perform, together with the management of the Horse in general.

Causes.—High feed and want of proper exercise, extensive wounds (and particularly of the joints); exposure to cold when in a state of perspiration; sudden transition from a cold to a hot temperature. Horses taken from grass and put suddenly into warm stables are extremely liable to those internal inflammations on which symptomatic fever depends.

Symptoms.—The symptomatic fever has many symptoms in common with simple fever, which are: loss of appetite, quick pulsed, dejected appearance, hot, mouth and debility. (If inflammation of the lungs is the cause of the fever, the designating symptoms will be, difficulty of breathing and a quick working of the flanks.) If the horse hang down his head in the manger, or lean back upon his collar with a strong appearance of being drowsy, the eyes appearing watery and inflamed, the fever depends upon an accumulation of blood in the vessels of the brain, and inflammation of the brain is the cause of the fever. In this case the pulse is not always quick; sometimes, indeed, I have found it unusually slow.

When the symptoms of fever are joined with yellowness of the eyes and mouth, an inflammation of the liver is indicated. Should an inflammation of the bowels be the cause, the horse is violently griped. An inflammation of the kidneys will also produce fever, and is distinguished by a suppression of urine and an inability to bear pressure upon the loins. When inflammation of the bladder is the cause, the horse is frequently stalling, voiding only very small quantities of urine, and that with considerable pain. When extensive wounds (and particularly those of the joints) is the cause of the fever, the cause cannot be mistaken. As symptomatic fever is caused by local inflammation of various organs, you will have the treatment of symptomatic fevers under their proper heads. As all fevers have many symptoms in common with each other, I have given you the designating symptoms in each disease, which will be of great use to the practitioner in discriminating disease.

Having now given a general description of symptomatic fever, I shall proceed to treat of those cases separately to which, above, I have briefly alluded.

DISEASES OF THE LUNGS.

INFLAMMATION OF THE LUNGS.

This disease requires the most prompt and energetic treatment; for so rapid is its progress in the Horse, that, unless checked at an early period, it generally proves fatal.

Symptoms.—Shivering; quickness of breath; the pulse is unusually quick, beating from sixty to eighty in a minute, whereas in health it is only about forty. The pulse is oppressed and indistinct; the artery is plainly to be felt under the finger, and of its usual size, but the pulse no longer indicates the expansion of the vessel, as it yields to the gush of blood, and its contraction when the blood has passed; it is rather a vibration or thrill communicated to a fluid already over-distending the artery. In a few cases even this almost eludes the most delicate touch, and scarcely any pulsation is to be detected. The sensation given to the finger, above described, is one of the plainest symptoms that we have to designate inflammation of the lungs from any other disease. Owing to the ill-decarbonized blood in the lungs, it cannot flow through the small or capillary vessels of the arteries of the lungs, which give this vibration or thrill. In addition to these symptoms, we have all the symptoms attending inflammatory disease in general.

Treatment.—The treatment depends wholly on the cause that produced the disease. If it arises from a plethoric or full habit, cathartics should be freely used. The Cathartic Powder No. 10 would be the most effectual. Dose : one ounce, to be repeated once in six hours, in a pint of gruel, until it operates as physic. This not only operates on the bowels, but produces its relaxing and expectorating effect on the lungs, and relieves them by exciting the mucous membrane to expel and throw off the adhesive mucus, which, if suffered to remain, would plug up the air-passages and cause hepatization. The operation of physic should be assisted by injections, composed of warm water and salt, frequently given. If the disease is caused by taking cold, producing a check of the perspiration, give the Sudorific Tincture No. 1, an ounce once in six hours, in a pint of flaxseed tea or thin gruel, and proceed as directed under *Mode of Producing Perspiration.* After the operation of the physic, or sweating, as the case may be, give two ounces of the Expectorant Powder No. 4, once in four or six hours, in a drench of a pint of flaxseed tea or gruel; the sides of the chest should be frequently rubbed with the Laxative Liniment No. 1 or 4. I have succeeded of late admirably well by the use of *veratrum viride* in all inflammatory diseases. Dose: from ten to twenty drops of the fluid extract, to be given in a drench of one pint of cold water once in six hours. This medicine has no equal in allaying arterial excitement and reducing inflammatory action ; and I prefer it to nauseating medicines of any kind.

The horse should be warmly clothed, that the blood may freely circulate through the capillary vessels of the skin; by so doing, we relieve the inflamed and over-distended internal organs. It is absurd to shut up every door and window ; the horse should have free access to pure air. It is interesting to see how eagerly the horse avails himself of the relief which this affords ; but no direct draft should blow upon him. It is indispensable that the legs should be frequently hand-rubbed to restore the circulation in them ; and if they should be inclined to be cold, they should be covered with thick flannel bandages. As to food, if the horse inclines to eat, a small quantity, hay, bran-mashes, carrots, &c., may be given.

After the inflammation has been subdued, the practitioner should not be in too great a hurry for the recovery of the patient, and commence the use of tonics too soon, as it is very easy to bring back the malady in all its violence. Nature will slowly, but surely, restore the appetite and strength.

CATARRHAL INFLAMMATION OF THE LUNGS.

Definition.—Catarrhal inflammation of the lungs differs from that I have before described, in being accompanied with a sore throat, difficulty of swallowing, weak cough, and a tendency to discharge from the nose. The pulse at the commencement of the disease is not very quick—sometimes not more frequent than in health; but it is generally weak, and not readily felt.

This disease generally prevails epidemically, and principally affects the membrane that lines the throat and windpipe. Nature often relieves the parts by throwing off a copious discharge of mucus, which is discharged through the nostrils.

Treatment.—First, Equalize the circulation and restore the function of the skin, as directed in article *Mode of Producing Perspiration*. Second, Relieve the soreness of the throat by the Stimulating Liniment No. 3, to be well rubbed on the parts morning and evening; fasten a flannel bandage around the neck. Third, Lubricate the mucous surface and regulate the secretion, and assist nature in discharging from the nostrils the thick and tenacious matter by steaming the head over the following decoction:—

Lobelia herb,	2 ounces;
Licorice,	2 "
Hops,	1 "

Steep in a pint of vinegar; pour the mixture, a little at a time, upon a red-hot brick held under the nose, several times a day, five minutes at a time.

If there is much difficulty in swallowing, give 20 drops of the tincture of belladonna, once in three hours, in a pint of water, until relief is obtained. If the fever runs high, give 30 drops of the fluid extract of veratrum viride, once in six hours, until the fever subsides. The after treatment should be the same as inflammation of the lungs.

CATARRH, OR COLD.

It would be useless to give a particular description of this complaint, since it is so generally understood. Scarcely any one can be at a loss to distinguish it from other disease. It consists in an inflammation of the mucous membrane which lines the internal parts of the throat, nose, &c. Sometimes it is attended with a slight degree of fever; hence arise the cough and discharge from the nostrils, which are the principal symptoms of catarrh. In slight attacks, all that is necessary is hot bran mash, frequently given, which will not only serve to keep the bowels open, but will act as a fomentation to the inflamed membranes, for the horse will be constantly inhaling the vapor which escapes from them. If there is much soreness of the throat and swelling of the glands, accompanied with fever, treat the case as catarrhal inflammation of the lungs. If a cough remains, you have the remedy in the article *Chronic Cough*.

STRANGLES, OR HORSE DISTEMPER.

The Strangles, as the name imports, is first indicated by a cough and difficulty of swallowing, as if the animal would die of strangulation. It is inherent in the nature of the Horse, and it is generally believed that not one escapes the disease. It is a disease incident to the young animal; that is, from two to three years old. I have always considered it as a critical disease, and treated it as such, encouraging the formation of matter, and assisting nature in throwing off a specific poison that is obnoxious to the constitution.

Symptoms.—A swelling commences between the jaw bones, directly under the tongue; great heat, pain, and tension of the tumors, and of all the adjacent membranes to such a degree that the animal can scarcely swallow. A discharge of thick, white matter from the nostrils follows. The eyes send forth a watery humor, and the animal nearly closes the lids. This is mostly the case when the two large glands under the ears are affected also, which frequently happens. The swelling increases, and bursts of itself.

Treatment.—The treatment of Strangles is very simple. As the essence of the disease consists in the formation and suppuration of the tumor under the jaw, the principal or almost the sole attention of the practitioner should be directed to the hastening of the process; therefore, as soon as the tumor of Strangles appears, the Stimulating Liniment No. 3 should be applied often. After the application of the liniment, a poultice should be put on, and frequently changed. As soon as the swelling is soft on the top, and evidently contains matter, it should be deeply and freely lanced. It is a bad, although frequent practice, to suffer the tumor to burst naturally, by which a ragged ulcer is formed, very slow to heal, and difficult of treatment.

The remainder of the treatment will depend on the symptoms. If there is much fever and evident affection of the chest, and which should carefully be distinguished from the oppression and choking occasioned by the pressure of the tumor, the treatment should be the same as in catarrhal inflammation of the lungs.

CHRONIC COUGH.

It would occupy more space than I can devote to this part of the subject to speak of all the causes of obstinate cough. Irritability of the air-passages, occasioned by previous and violent inflammation of them, is the most frequent. It is sometimes caused by worms. There is much sympathy between the lungs and the intestines, and the one very readily participates in the irritation produced by the other. It is the necessary attendant of thick wind and broken wind, for these proceed from alteration in the structure of the lungs.

Treatment.—Notwithstanding the clearness of the cause, the cure is not so evident. If a harsh hollow cough be accompanied by a staring coat, and the appearance of worms, the Vermifuge No. 3 may expel the worms, and remove the irritation of the intestinal canal. If it proceeds from irritability of the air-passages, which will be discovered by the horse coughing after drinking, or when he first goes out of the stable, in the morning, or by his occasional throwing out thick mucus from the nose, medicines should be given to diminish irritation of the lungs. The Anodyne Powder, No. 6, is very effectual, given three times a day in his feed. If this does not succeed, give the Expectorant Powder three times a day. Feeding has much influence on this complaint; all dry food should be avoided; the feed should consist of bright oat-straw, oats perfectly clear from dust, carrots, &c. Carrots afford decided relief. When coughing chiefly occurs after eating, the seat of the disease is evidently in the substance of the lungs. The stomach, distended with food, presses upon the diaphragm, the diaphragm upon the lungs, and the lungs, already laboring under some congestion, are less capable of transmitting the air. In the violent efforts to discharge their function, irritation is produced, and the act of coughing is the consequence of that irritation; this soon runs into a disease called

THICK WIND.

Symptoms.—A short, frequent, and laborious breathing, and especially when the animal is in exercise; inspiration and expiration often succeeding each other so rapidly as evidently to cause distress, and occasionally almost to cause suffocation.

Cause.—The principal cause of thick wind is previous inflammation, and particularly inflammation of the branches of the lungs. The throwing out of some fluid, which is capable of coagulation, is the result, or the natural termination of inflammation. This deposit in the substance of the lungs, or in the bronchial tubes, from inflammation of these organs, must close many of the air-cells, and lessen the dimensions of others. Then, if the cells, fewer in number and contracted in size, be left for the purpose of breathing, the rapid and laborious action of the lungs must supply the deficiency, and especially when the animal is put in that state in which he requires a rapid change of blood. The examination of thick-winded horses has thrown considerable light on the nature of the disease. In the majority of instances some of the small air-cells have been found filled with a dense substance of a blue or dark color; in others, the minute passages leading to the cells have been diminished and almost obliterated, the linings of these passages being unnaturally thickened, or covered with hardened mucus; and where neither of the appearances could be found, the lining of the cells has exhibited evident marks of inflammation, so that pain prevented the full expansion of the lungs.

Thick wind is often the cause of broken wind. It is easy to understand this; for if so much labor is necessary to contract the air-cells and force out the wind, and the lungs work so rapidly and so violently in effecting this, some of the cells, weakened by disease, will be ruptured.

Treatment.—This depends more upon management than medicine. Attention to diet, and the prevention of overloading the stomach, and the avoidance of exercise soon after a meal, may in some degree palliate the disease. All dust from hay or oats should be avoided. The horse should have constant exercise, carried to the extent of his powers, without too much distressing him. The capability of exertion will thus daily improve, and the breathing of the horse will become freer and deeper. In this consists the secret of training the horse for the course; and this constitutes all the difference between a horse that has been well and one that has been badly trained.

Thick wind and broken wind exist in various degrees, and many shades of difference, which have been denominated by different names among horsemen, such as *Piping, Roaring, Wheezing, Whistling, High-Blowing, Grunting,* &c.; names very expressive, though they can boast no elegance.

Pipers may be known by a shrill noise made by the horse when in quick action. This is usually caused by a ring of coagulated matter forming round the inside of the windpipe; the cavity, in consequence, is naturally diminished, and the sound produced in quick breathing must evidently be more shrill.

Wheezing is caused by the lodgment of mucus in the small passages of the lungs; it frequently accompanies bronchitis. Sometimes wheezing is produced by a contraction of the small passages of the lungs, which utter a sound like that of an asthmatic person after a little over-action.

Roaring.—This disease takes its name from a peculiar sound in respiration, particularly when the horse is put into brisk motion. It is caused by a lymph which has been collected in the windpipe or its branches, which, becoming solid, obstructs, in a greater or less degree, the passages of air.

Whistling.—The whistler utters a shriller sound than the wheezer. The disease seems to be referable to some contraction in the windpipe or larynx.

High-Blowing.—This appears to be an obstruction, principally in the nose; the horse loudly puffs and blows, and the nostrils are distended to the utmost, while the flanks are comparatively quiet.

Grunting.—Every horse violently exercised on a full stomach, or when overloaded with fat, will grunt very much like a hog. It is caused by the relaxation of the midriff which is produced by this distension of the stomach; consequently, as the horse strikes with his fore feet when moving, the stomach is forced upon the lungs; this forced expiration will be accompanied by this kind of sound.

HEAVES.

Generally termed broken wind. As to the cause, symptoms, and general appearance of the lungs in this disease, there have been various opinions among veterinary surgeons; their theories have generally been conflicting; they have had controversies without end. Richard Lawrence, veterinary surgeon, says: "The most common appearance in the lungs of broken-winded horses is a general thickening of their substance, by which their elasticity is in a great measure destroyed, and their weight specifically increased; at the same time, their capacity for air is diminished. During life, the lungs entirely fill the cavity of the chest, so as to leave no space between their outward surface and the inward surface of the ribs; thus they dilate and contract by their own elasticity, and the action of the ribs and diaphragm. If the chest be punctured in the dead subject, the air rushes in and the lungs collapse; but if the horse were broken-winded, the lungs do not collapse. This state of the lungs sufficiently accounts for the difficulty of dilation. The ribs cannot expand without forming a vacuum in the chest, which the pressure of the external air prevents. On this account the air is received into the lungs with difficulty; but its expulsion is not so difficult, as the return of the ribs and diaphragm naturally force it out by their pressure. Thus in broken-winded horses, inspiration is very slow, but expiration is sudden and rapid, as may be seen by the flanks returning with a jerk."

White, having taken up Mr. Lawrence rather sharply, and somewhat unjustly, says: "It appears to me that the observations of Mr. Lawrence on this subject are not correct; the lungs of broken-winded horses which I have examined, have been unusually large, with numerous bladders on the surface. This must have arisen from a rupture of some of the air cells; for in this case, some part of the air which is inspired will necessarily get into the cellular membrane of the lungs, and diffuse itself until it arrives at the surface, when it will raise the pleura so as to form the air-bladders we observed. This is the reason that the lungs of broken-winded horses do not collapse when the chest is punctured; and this will serve to explain the peculiar motion of the flanks in broken-winded horses, which does not consist, as Mr. Lawrence asserts, in a quick expiration and a very slow inspiration, but quite the reverse. Air is received into the lungs very readily, which is manifested by a sudden falling of the flanks; but its expulsion is slow and difficult, as may be seen by the long-continued exertions of the abdominal muscles."

I am, however, fully convinced of the cause of this diversity of opinion respecting the precise condition and appearance of the lungs of broken-winded horses. Frequent dissections have convinced me that there are three different kinds of heaves. The lungs present a different appearance in each, owing to the nature of the disease, and the cause that produced it. If caused by inflammation and consequent thickening of the substance of the lungs, they are found in precisely the same condition as described by Lawrence; but if caused without previous disease, and in consequnence of a distension of the stomach and rupture of the air cells, the lungs will be found in the same condition as described by White. Therefore, the different opinions arise from examining the lungs of horses under different kinds of heaves. These three kinds of heaves may be designated by the manner of breathing. If caused by rupture of the air cells, the air will rush in easily enough, and one effort of the muscles of respiration is sufficient for the purpose; but when the air gets into the cellular membranes of the lungs in consequence of a rupture of the cells, there is no regular outlet, and the air is as readily forced backward as forward, and it is exceedingly difficult to force it out again; and two efforts are scarcely sufficient. Thus inspiration is quicker than expiration, as described by White.

In that kind of heaves accompanying the thickening of the membrane, as described by Lawrence, the manner of breathing is directly the reverse—inspiration is slow, and expiration is sudden and rapid, and especially when accompanied by a thickening of the midriff, or its adhesion to the stomach. The third kind is a spasmodic action, in which inspiration and expiration are equally difficult, and occupy precisely the same time. At times the breathing is exceedingly difficult—at other times, the patient appears very much relieved.

Cause.—We do not find heavey horses on the race-course, for, although every exertion of speed is required, their food lies in a small compass; the stomach is not distended, and the lungs have room to play. Carriage and coach-horses are seldom heavey, unless they bring the disease to their work, for they, too, principally live on grain, and care is taken that they are not fed immediately before they are set to work. The majority of heavey horses come from the stable of the farmer, for whose use these pages are principally designed. The way they generally manage the colt, when they take it from the company of the mare, is to turn it into a distant pasture to shift for itself, or put it upon dry hay, with water, perhaps, once or twice a day. The colt, in order to get sufficient nourishment to supply the demands of nature, is compelled to take in large quantities of this coarse, innutritive food; this, together with the large quantity of water required to digest it, distends the stomach to an enormous size; the midriff, in consequence of this, becomes relaxed and forced in against the lungs; this prevents their full expansion,—they are thereby prevented from inspiring a sufficient quantity of air to fill them, when a portion of the air-cells, not distended with air, fill up or contract, and become useless. There will be but few horses managed in this way from colts, but what will have the heaves. At least, all the best of them will be thus affected; such as are hearty eaters, and keep themselves in good condition on hay; these are large-bellied horses in consequence of their hearty eating, and are almost always heavey horses. On the other hand, narrow-chested, slim horses seldom have the heaves, and will never keep in good condition on hay; and never keep so full as to hinder the free action of the lungs. This is most conclusive proof that heaves is generally brought on by keeping the horse on coarse, bulky food. Experience has taught me that if the farmer would feed his horse more grain and less hay, there would be few or no cases of heaves.

The reason why there are more heaves now than heretofore under the same management is, because it has become hereditary—like will produce like—the progeny will inherit the qualities or mingled qualities of the parents. The colt will inherit weak lungs, or heaves, or a predisposition to it.

As to the treatment for heaves, it depends more on management than it does on medicine. It is vain to attempt to cure disease as long as the cause remains in full effect. The lungs were originally, and in a state of nature, perfect, and this disease has been brought on by transgressing the laws of health of the lungs; so after disease has been cured, it is unreasonable to suppose it will not return again, if subject to the same cause that produced it. We must not expect that diseased lungs can be rendered more perfect by medicine than they were in a state of nature. Although the heaves is laid down by all authors as an incurable disease, yet it can and may be cured. I do not profess to have discovered a specific or a cure-all; neither do I make any medicine a favorite hobby, and give it without regard to pathological conditions; neither do I pretend that it is curable in all stages, yet the majority of cases may be cured, and probably all might have been cured, if attended to at an early period, and all cases admit of relief.

Treatment.—The food of the animal should consist of much nutriment condensed into a small compass,—the quantity of oats should be increased, and that of hay diminished, or they should have no hay at all; instead of hay, give good bright straw, and keep the horse clear from dust, and give the Expectorant Powder No. 4, morning and evening, and I will risk the heaves. You will find many valuable receipts for the heaves under the head Receipts.

DISEASES OF THE STOMACH.
INFLAMMATION OF THE STOMACH.

The stomach, like the intestines, may be inflamed either on the external or internal surface.

Symptoms.—When the external coat is the seat of the disease, the symptoms are nearly the same as those by which peritoneal inflammation of the intestines is indicated; the only difference observable in the symptoms is, that in this case the pain seems to be more acute and distressing than in the other. The same difference may be observed between the large and small intestines, the latter being possessed of more sensibility than the former.

An inflammation of the internal or villous coat of the stomach is not very common, and is generally occasioned by poisons or strong medicines that have been given, or by a species of worms termed bots, of which you will have the treatment in their proper place. If it does occur independent of the above causes, the treatment consists in giving oily or mucilaginous liquids freely, such as decoction of linseed, gum-arabic dissolved in water, &c. When the inflammation is confined to the peritoneal or external coats of the stomach, apply the Stimulating Liniment No. 3 on the chest freely, and give the

Diaphoretic Powder morning and evening. If the extremities are cold, manage as directed in article *Mode of Producing Perspiration.* Clysters are to be injected, and, if the disease be accompanied with purging, they should be composed of strong linseed decoction or water-gruel.

STOMACH STAGGERS.

There are two varieties of this disease—the sleepy or stomach, and the mad staggers; frequently, however, they are only different stages of the same disease, or varying with the cause that produced them.

Symptoms.—In stomach staggers the horse appears dull, sleepy, staggering; when roused, he looks vacantly around him, seizes a lock of hay, and dozes again with it in his mouth. At length he drops and dies, or the sleepiness passes off and delirium comes on, when he falls, rises again, drops, beats himself about, and dies in convulsions.

Cause.—The cause of this disease is sufficiently evident, and the disease never occurs except by mismanagement of the horse. It arises from over-feeding and want of proper exercise, or by getting at a great quantity of food of an improper nature, or by keeping the horse too long without eating. When he has been hard at work, and has become very hungry, he falls ravenously upon every kind of food he can get at, swallowing it faster than his small stomach can digest it; and no water being given to soften it and to hasten its passage, the stomach becomes crammed, and having been previously exhausted by long fasting, is unable to contract upon its contents; the food soon begins to ferment and to swell, causing great distension; the brain sympathizes with this over-loaded organ, and staggers are produced. We can easily imagine this when we remember the sad headache occasioned by over-loading the stomach. Sometimes the horse's stomach is ruptured by eating large quantities of dry food previous to being watered.

Treatment.—We should be most diligent and minute in our inquiry into the history of the horse for the preceding twenty-four hours, whether he could have got an undue quantity of food, or had been worked hard and kept long fasting; for it is almost or quite impossible for the person most accustomed to the horse to distinguish between the early stages of stomach and mad staggers (distension of the stomach and inflammation of the brain). If the disease is caused by a distension of the stomach and a fermentation of its contents, which throws out an acid, the first thing to be done is to neutralize the acid by giving

   Soda, ... 4 ounces,
   Water, ... 1 quart,

to be given as a drench. After the neutralization of the acid, give the Cathartic Powder No. 2, at intervals of six hours, until you get an operation; if one ounce of the Stimulating Powder No. 1 be added to the above, it would assist the operation of the physic by stimulating the stomach and bowels to action. Let the hand be introduced into the anus, and remove all the hardened dung that can be found. The following injection is to be given:—

   Common salt, ... 8 ounces.
   Warm water, ... 1 gallon.
   Olive oil, ... 4 ounces.

This clyster should be repeated often until you get an operation of physic.

After the operation of physic, no further medication is necessary. Care should be taken that the food should be light and of easy digestion.

DISEASES OF THE BOWELS.

INFLAMMATION OF THE BOWELS.

Inflammation may affect the internal or mucous coats, or the external or peritoneal coats. The first is accompanied by violent purging, and is usually the consequence of drastics or an over-dose of physic. The second is accompanied by considerable fever and costiveness. Inflammation of the external coat of the bowels is a very frequent and fatal disease. It speedily runs its course, and it is of great consequence that its early symptoms should be known.

Symptoms.—If the horse be carefully observed, restlessness and fever will have been seen to precede the attack; in many cases a direct shivering fit will be observed; the mouth will be hot, and the nose red. The horse will soon express the most dreadful pain, by pawing, striking at his belly, looking wildly at his flanks, groaning and rolling; the pulse will be quick and small, the ears and legs cold, the belly tender and hot, the breathing quickened, the bowels costive, and the horse becoming rapidly and fearfully weak.

Inflammation of the bowels may be distinguished from colic by the following symptoms :—The disease will be gradual in its approach, with previous indications of fever, pulse quick and small, legs and ears cold, belly hot and tender, motion increases the pain, constant pain, rapid and great weakness. Colic is sudden in its attack, pulse regular, legs and ears of the natural temperature, relief obtained by rubbing the belly, or by motion, intervals of rest, strength scarcely affected.

Cause.—Sudden exposure to cold when in a state of perspiration, drinking freely of cold water when too warm, being drenched with rain, or having the belly washed with cold water, &c.

Treatment.—Mucilaginous drinks, such as gum arabic, slippery elm, flaxseed, flour porridge, &c., should be freely given. Although the bowels are constipated, drastic cathartics should be avoided. I have learned by experience that cream of tartar and sulphur are the most proper in this disease.

Cream of tartar,	2 ounces,
Sulphur	2 "

to be given in one pint of thin gruel, to be repeated in six hours, if it does not sooner operate.

The operation should be assisted by the frequent use of clysters, composed of—

Linseed oil,	1 pint,
Lime water,	1 qt.

Counter-irritants are of great use. The following I have found to be the most effectual, owing to its relaxing and stimulating properties :—

Powdered Tobacco,	2 ounces,
" Lobelia seed,	1 "
" Capsicum,	1 "
" Slippery elm,	2 "
Soft soap,	1 pint.

Mix well. The whole belly should be covered with the above mixture, and well rubbed in.

No corn or hay should be given during the disease, but bran-mashes, and green food, if it can be procured. The latter will be the best of all food, and may be given without danger.

Inflammation of the mucous membrane of the bowels is generally caused by physic, given in too great quantity, or of a drastic kind. The purging is more violent, and of longer continuance than was intended; the animal shows that he is suffering great pain; he frequently looks round at his flanks; his breathing is laborious.

Treatment.—We should plentifully administer gruel, or thin starch, or arrowroot, by the mouth and by clysters; also give

Paregoric, . . . 4 ounces.

once in six hours. As soon as the purging begins to subside the paregoric should be lessened in quantity, and gradually discontinued. It sometimes happens that the paregoric is not sufficient to stop purging; in this case give

Laudanum, . . . 1 ounce,
Tincture Kino, . . 1 "
Peppermint essence, . . 2 "

to be given in a pint of gruel.

Violent purging, attended with much inflammation and fever, will sometimes occur from other causes. Green food will sometimes purge; a horse worked hard upon green food will scour. The remedy is change of diet, or less labor. If a change of diet is not sufficient to cure the disease, give the the above treatment.

SPASMODIC COLIC.

The passage of food through the intestines is effected by the alternative contraction and relaxation of the musular coat of the intestines. When the action is simply increased through the whole of the canal, the food passes more rapidly, and purging is produced; but the muscles of every part of the frame are liable to irregular or spasmodic action, and the muscular coat of some portion of the intestines may be thus affected. A species of cramp may attack a portion of the intestines. This spasm may be confined to a small part of the canal. In the Horse, the ilium is the usual seat of the disease. The gut has been found after death strangely contracted in various places—contraction not extending above five or six inches in any of them. It is of much importance to distinguish between spasmodic colic and inflammation of the bowels, for the symptoms have considerable resemblance, although the treatment should be very different.

Symptoms.—The attack of colic is usually very sudden. There is often not the slightest warning. The horse begins to shift his position, look round at his flanks, paw, violently strike his belly with his feet, lie down, roll, and frequently on his back. In a few minutes the pain seems to cease, the horse shakes himself, and begins to feed; but on a sudden the spasm returns more violently; every indication of pain is increased; he heaves at the flanks, breaks out into a profuse perspiration, and throws himself more violently about. In the space of an hour or two, either the spasms begin to relax, and the remissions are longer in duration, or the torture is augmented at every paroxysm, the intervals are fewer and less marked, and inflammation and death supervene.

Cause.—Drinking of cold water when the horse is heated. There is not a surer cause of violent spasm than this. The exposure of a horse to cold air or cold wind after violent exercise, green food in too large quantities when the horse is hot. In some horses there seems to be a constitutional predisposition to colic. They cannot be hardly worked, or exposed to unusual cold without a fit of it.

Cure.—Although this disease is a frequent one, and often proves fatal, fortunately we have found a medicine that allays the spasms, and the disease often ceases almost as suddenly as it appears. Laudanum is one of the most powerful remedies, and especially if given in combination with turpentine.

Spirits of turpentine, . 3 ounces;
Laudanum, . . . 1 "

given in a pint of starch water, will frequently give almost instantaneous relief. I have found the Cholera Tincture No. 6, to be very effectual, given in one-ounce doses, in a pint of warm water or gruel. For want of these articles, give black pepper, ginger, cloves, or capsicum. Gin and black pepper is a good medicine. The belly should be rubbed well with pepper-sauce, red pepper and vinegar, or some other stimulant. The horse should be walked about, or trotted moderately. The motion thus produced in the bowels, and the friction of the intestines, may relax the spasm; but a hasty gallop may speedily cause inflammation to succeed to colic. Clysters are of the utmost importance, and will frequently cure of themselves. They may be composed of salt and water, or simply warm water, and should be frequently injected.

When relief has been obtained, the horse, saturated with perspiration, should be well clothed; he should be well littered down in a warm box or stall, have bran-mashes for the next two or three days, and drink lukewarm water.

WORMS.

Different kinds of worms inhabit the intestines of the Horse, but, except when they exist in great numbers, they are not so hurtful as is generally supposed, although the groom may trace to them hide-binding, cough, loss of appetite, gripes, magrums, and a variety of other ailments. Of the origin or mode of propagation of these parasitical animals we will say nothing; neither writers on medicine, nor even on natural history, have given us any satisfactory account of the matter.

The long white worm, *lumbricus tires*, much resembling the common earth-worm, and being from six to ten inches long, inhabits the small intestines, and, if there are many of them, they may consume more than can be spared of the nutritive part of the food or the mucus of the bowels.

A smaller dark-colored worm, called the needle-worm or ascaris, inhabits the large intestines. Hundreds of them sometimes descend into the rectum, and immense quantities have been found in the cœcum. They are a more serious nuisance than the former, for they cause a very troublesome irritation about the fundament, which sometimes sadly annoys the horse. Their existence can generally be discovered by a small portion of mucus, which, hardening, is converted into a powder, and is found about the anus.

Treatment.—A dose of physic will sometimes bring away almost incredible quantities of them. When a horse can be spared, a strong dose of physic is an excellent vermifuge, so far as the long round worm is concerned; but perhaps a better medicine, and not interfering either with the feeding or work of the horse. is the Vermifuge No. 1. For the small dark-colored worm, when they descend into the rectum or lower part of the bowels, use a clyster of a strong infusion of tobacco, or linseed oil, or aloes dissolved in warm water, and freely given.

BOTS.

Horses are much troubled by a grub or caterpillar, in the early part of the spring, which crawls out of the anus and fastens itself under the tail, and seems to cause a great deal of itching or uneasiness. Grooms are sometimes alarmed at their appearance. Their history is curious, and will dispel every fear in regard to them.

A species of fly called the gad-fly is, in the latter part of summer, exceedingly busy about the horse. They are observed to be darting with great rapidity about the knees and sides of the animal. The female is depositing the eggs on the hair, which adhere to it by means of a glutinous fluid with which they are surrounded. In a few days the eggs are ready to be hatched; and the slightest application of warmth and moisture will liberate the little animals which they contain. The horse, in licking himself, touches the egg; it bursts, and a small worm escapes, which adheres to the tongue, and is conveyed with the food into the stomach; there it clings, by means of a hook on either side of its mouth, to the cuticular portion of the stomach, and its hold is so firm and so obstinate that it will be broken before it will be detached. It remains there, feeding on the mucus of the stomach during the whole of the winter and to the end of the ensuing spring; when, having attained a considerable size, and being destined to undergo a certain transformation, it is disengaged from the cuticular coat of the stomach, and is carried off through the intestines by their peristaltic action and expelled with the dung. After the larva or maggot is thus expelled, it seeks shelter in the ground, contracts in size, and becomes a chrysalis grub; in which state it lies inactive for a few weeks, and then, bursting from its confinement, assumes the form of a fly. The female, after becoming inpregnated, quickly deposits her eggs on those parts of the horse which he is most likely to lick, and so the species is perpetuated.

The conclusion drawn from the history of this grub is, that it cannot, while it inhabits the stomach of the horse, give the animal any pain, for it fastens only on the cuticular and insensible coats. They cannot be injurious to the horse, for he enjoys the most perfect health, when the cuticular part of the stomach is filled with them and their presence is not even suspected until they appear at the anus. They cannot be removed by medicine, because they are not in that part of the stomach to which medicine is usually conveyed; and if they were, their mouths are too deeply buried in the mucus for any medicine, that is safe to be administered, to affect them. It is a common belief that bots bore through the walls of the stomach. This, I contend, is never the case while the horse is alive. Instinct teaches this little animal not to attempt to leave its comfortable and natural location for one that is fatal to his existence. When the bot has fulfilled his mission in the horse's stomach, and becomes fully developed as a grub, his instinct teaches him to leave for a more congenial clime, where he can undergo a transformation to the fly again. This is as natural a process as that an egg should produce a chicken, and the chicken a hen; and is governed by the same universal law. When the bot detaches itself from the stomach, it is carried off by the peristaltic motion of the alimentary canal. I do not deny that bots are found in the abdominal cavity, for as soon as the horse dies, the animal is subject to the laws of decomposition; and what had previously been food for the bots is now their poison, and they must themselves become subject to this chemical action and be destroyed, or escape from it. The peristaltic action of the intestines having ceased at death, the bots cannot pass through their natural outlet; their nature suggests to them that the only means of escape would be through the walls of the stomach, which show but little resistance, being partly decomposed; they easily perforate the stomach, and are found in the abdominal cavity. Therefore, the wise man will leave them to themselves, or content himself with picking them off when collected under the tail of the animal.

After becoming acquainted with the history and mode of propagation of the bot, I was convinced in my mind that the bots never killed the horse by eating through the stomach or maw. So firm was my convictions that I purchased an old horse to try the experiment; eight hours after the horse was killed, I made an examination, and found several bots in the abdominal cavity, that had perforated the stomach of the animal. I knew that the horse did not die of the bots, although this would have been the decision of ninety-nine out of a hundred that did not know the cause.

Every man that chances to come along will be a farrier, and will say your horse has the bots, no mistake, and will have a specific, when he is as ignorant of the nature and history of the bot as a toad is of a pocket-book.

INFLAMMATION OF THE LIVER.

The Horse not being exposed to the causes which produce this complaint in man, it is consequently a disease of rare occurrence in this animal. Although his food is sometimes highly nutritive, the work which he is compelled to perform prevents it from unduly stimulating this important organ; and when inflammation of the liver does occur, it is difficult to distinguish it from that of the bowels; so much so, that even the professional man is liable to be deceived.

Symptoms.—Yellowness of the skin and eyes are the only designating symptoms. The disease may be known by red and dark-colored urine; great weakness and fever, sometimes accompanied with costiveness, but generally with diarrhœa or purging. The horse has a very languid appearance, and is almost constantly lying down. Sometimes the progress of this disease is very rapid, speedily terminating in death; at other times it proceeds more slowly, and in this case it frequently terminates in dropsy or inflammation of the bowels.

Treatment.—If costiveness is one of the symptoms, the following ball may be given:

Castile soap, ... 3 drachms.
Barbadoes aloes, . . . 1 "
Rhubarb, ... 4 "
Syrup, enough to form a ball for one dose.

To be given once in six hours, until it occasions moderate purging; but if the bowels are in a laxed state, or attended with diarrhœa give the following:—

Gum kino, ... 1 drachm.
Opium, ... 1 "
Prepared chalk, ... 2 "

This may be given in powder, or made in a ball by the addition of a little syrup, to be given at a dose. The Stimulating Liniment should be used as directed in inflammation of the bowels.

The strength of the horse should be promoted; the diet should be nutritious and of easy digestion, such as bran, carrots, arrow-root, &c. After the inflammatory symptoms have passed off, give

   Unicorn, ... 1 ounce,
   Golden seal, ... ½ "

morning and evening; this will operate as a tonic and alterative.

JAUNDICE.

Jaundice is caused by some obstruction in the ducts or tubes which convey the bile from the liver to the intestines, or it may be caused by a torpid, inactive state of the liver. The horse has no gall-bladder, in which it can become thickened, and harden into masses and produce gall-stone, which is a very common disease in the human subject.

Symptoms.—Yellowness of the eyes and mouth; urine high-colored; the horse is languid, and the appetite impaired; the dung is small and hard; this disease may be distinguished from inflammation of the liver by the absence of fever.

Treatment.—We must endeavor to restore the natural passage of the bile by alteratives, given in small quantities, repeated at short intervals, until the bowels are freely opened. For this purpose give the following powder :—

   Pulverized mandrake, . . 2 drachms.
   " bloodroot, . . 1 "
   " blue flag root, . ½ "

To be given at a dose. After the bowels are freely opened, the powder should be given once or twice a week for some time. To change the condition of the secretions, carrots or green food will be very beneficial. Tonics should be given, composed of

   Golden seal, ... 2 drachms.
   Ginger, . . . 2 "

 • to be given at a dose, once each day, until the horse completely recovers.

TUMORS ON THE LIVER.

The liver is also frequently affected with tumors on its fine surface, as well as with ulcers or scirrhus, which are the effects of an impure state of the blood, of over action, and probably of external injuries, &c.

We can easily conceive that the thin parts of this large viscus may be diseased and inflamed, without causing derangement of the biliary function, further than that by increasing its action, and thinning the blood over-much, it obtains more bile. The animal then grows thin, though devouring his food as usual for a while; but we may ascertain when this evil has begun by the state of his dung, principally as to color, which will then be of much darker color. As pale dung is a symptom of suppressed bile, so is deep-colored an indication of a superabundance, that is caused by over-action.

When the dung has a pale color, it would indicate alteratives to restore the action of the liver; for this purpose you will find the Cathartic Powder No. 2 admirably adapted, given in half or quarter doses. If tumors in the liver do exist, they admit of no remedy.

INFLAMMATION OF THE KIDNEYS.

Inflammation of the kidneys is one of the most common diseases of the horse, and is more fatally and unskillfully treated than any other to which he is liable. Owing to the heavy burthens he is compelled to carry, together with the immense weight he is capable of drawing, it is not surprising that strain of the back and inflammation of the kidneys are often the result.

Symptoms.—The first symptoms are those of fever generally, but the seat of the disease soon becomes evident. The horse stands with his hind legs wide apart, straddling as he walks; manifests pain in turning; shrinks when the loins are pressed; and some degree of heat is felt. The urine is voided in small quantities, and frequently it is high colored, and sometimes bloody. The attempt to urinate becomes more frequent, and the quantity voided smaller, until the animal strains painfully, and looks anxiously round at his flanks.

Causes.—Too powerful or too often-repeated diuretics are the most frequent and powerful means of producing the disease. Mow-burnt hay, or musty oats, or oats that have been dried on the kiln, acquire a diuretic property. And if horses are long fed on them, the continual excitement of this organ which they produce will cause inflammation. Sprain in the loins is often transferred to the kidneys, with which they lie in contact. Carrying a heavy burthen, or a heavy rider, and being urged on far or fast. Exposure to cold is another frequent cause of the malady, especially if the horse be drenched with rain, or wet dripping on his loins; and more particularly, if the horse were previously disposed to inflammation, or the kidneys had been previously weakened.

Treatment.—The treatment will only vary from that of inflammation of other organs by the consideration and peculiarity of the parts affected. This is a disease that I never have seen treated according to correct pathological principles. They all give diuretic medicines, which is as absurd in this disease as it would be to give drastic cathartics in inflammation of the mucous membrane of the bowels. Diuretics irritate the parts already too much excited, and add fuel to fire, and frequently destroy the animal. An active purge should be given which by exciting irritation in the intestines acts as a counter-irritation and relieves the diseased organ. For this purpose give the Cathartic Powder No. 2; if this does not operate in eight hours, repeat the dose. After the operation of the purgative give fluid extract of veratrum

Veratrum, . . . 40 drops.
Water, ... 1 pint.

To be given as a drench three times a day, until the inflammatory action subsides; then give

Tincture of colchicum, . . 1 drachm,
Water, ... 1 pint,

once a day until the horse is cured.

INFLAMMATION OF THE BLADDER.

Inflammation may affect either the neck or the body of the bladder. The symptoms are nearly the same with those of inflammation of the kidneys, except that there is rarely a total suppression of urine, and heat is felt in the rectum over the situation of the bladder.

Causes.—The causes are the presence of some acrid or irritant matter in the urine, or calculi, or stone in the bladder, or the free use of irritating diuretics.

Treatment.—The treatment will be nearly the same as in inflammation of the kidneys, except that it is of more consequence that the horse should drink freely of mucilages, as a solution of gum arabic, flaxseed tea, &c. In inflammation of the neck of the bladder, there is the same frequent voiding of urine in small quantities, often ending in total suppression. In inflammation of the neck of the bladder, the bladder is distended with urine, and may be distinctly felt by introducing the hand into the rectum. If a spasm of the part closing the neck of the bladder causes the distention, the object to be attempted is sufficiently plain. This spasm must be relaxed. The most likely means to effect this is to give the Cathartic Powder No. 2 in a pint of starch-water, once in twelve hours until you get an operation. The operation should be assisted by the free use of clysters of a mucilaginous nature. After the operation of physic, give the Expectorant Powder No. 4. This will relax the spasm by its nauseating and relaxing qualities. In the mare, the bladder may easily be evacuated by means of a catheter, in skillful hands; but owing to the curved direction of the penis, a catheter cannot be introduced into the bladder of the horse, without an operation to which a veterinary surgeon alone is competent.

DIABETES, OR PROFUSE STALING.

This is a rare disease in the Horse. It is generally caused by undue irritation of the kidneys, bad food, or strong diuretics, and sometimes follows inflammation of the kidneys.

Treatment.—It being caused by an increased action of the kidneys, the most rational plan of treatment is to endeavor to abate that action; for which purpose give

Wine of colchicum, . . 2 drachms;
Water, . . . 1 pint;

to be given as a drench once a day. If this does not succeed, use a strong infusion of princes' pine; or give

Rosin, ... 1 ounce,
Loaf sugar, ... 2 "

in his feed twice a day. Very careful attention should be paid to the food. The hay and oats should be of the best quality, and green food, especially carrots, will be very serviceable.

INFLAMMATION OF THE BRAIN.

This disease generally attacks horses that are highly fed and moderately worked. It does not occur frequently, and is difficult to cure unless attended to at an early period.

Symptoms.—Inflammation of the brain can, as I have said, be at first with difficulty distinguished from the sleepy or stomach staggers; but after a while the horse suddenly begins to heave at the flanks, his nostrils expand, his eyes unclose, he has a wild and vacant stare, and delirium comes on rapidly. This delirium continues either until his former stupor returns, or he has literally worn himself out in frightful struggles. The pupil of the eye will be very much contracted and the eyes will be half closed, so as to exclude the light. There are only two diseases with which it can be confounded, and from both it is very readily distinguished, viz.: colic and madness. In madness there may be more or less violence, there is sometimes a determination to do mischief; and there is always consciousness. In colic, the horse rises and falls, but not with so much violence; he sometimes plunges, but he more often rolls himself about; he looks frequently at his flanks with an expression of pain, and he is conscious.

Causes.—Inflammation of the brain is generally caused by over-exertion when the horse is too fat or too full of blood, or especially during hot weather; but what will produce general fever may be the cause of this disease.

Treatment.—Physic is one of the most important remedial agents that can be given in this disease, and it is often necessary to give the most drastic kind. I have succeeded admirably well with the following powder:—

  Pulverized aloes, . . 1 ounce.
  " mandrake, . . 2 drachms.
  " ginger, . . 1 "
  Water, ... 1 pint.

This should be given as a drench, and followed up once in six hours until you produce the desired effect. To assist the operation of the physic, injections should be given every four hours, composed of

  Powdered lobelia, . . 1 ounce.
  " blood-root, . . ½ "
  Hot water, . . . 2 quarts.

The head should be kept cool with water. Nauseating medicines given in injections will do as much to relieve congestion and inflammation as any other medicines, and never should be omitted. The feet and legs should be kept warm, as directed in article *Mode of Producing Perspiration.*

MEGRIMS.

This disease is generally produced by pressure on the brain, and is caused by some fluid thrown out between the membranes of the brain, and occupying and distending the ventricles of the brain, and a determination or flow of blood to it. This organ requires a large supply of blood to enable it to discharge its important functions. It is supposed that nine tenths of all the blood in the body flows through the brain. Nature, in the Horse, has made some admirable provisions to cause this great quantity of blood to flow into the brain without too much velocity, and thereby to lessen the risk of suddenly overloading it or rupturing its vessels. The arteries pursue their course to the brain in a strangely winding manner, and they enter the skull through bony holes, which will admit of the enlargement of the arteries only to a very limited extent, yet from various causes the blood is forced to the head with sufficient force to produce the above-named disease.

Causes.—The most common, is violent exercise on a hot day. When the horse is fat and full of blood, more than the usual quantity is sent to the head; or from a small collar, or the curb-rein being too tight, the blood will be prevented from returning from the head, and thus the large vessels of the brain will be too long and injuriously distended, and the small vessels that run through the substance of the brain will be enlarged, and the bulk of the brain increased, and press upon the nerves, and produce the Megrims.

Symptoms.—Sometimes the horse will be going along regularly, when, all at once, he will stop, shake his head, and be evidently giddy, and half unconscious. In a minute or two this will pass off, and he will go on again as if nothing had happened. Frequently, however, the attack will be of a more serious nature. He will fall without the slightest warning, or suddenly run round once or twice, and then fall. Sometimes he will lie in a state of complete insensibility; at other times he will struggle with the utmost violence. In a few minutes he will begin to gradually come to himself; he will get up and appear quite regular, and proceed on his journey, yet somewhat dull, and evidently exhausted by what has happened, although not seriously ill.

This is a very dangerous disease both to the rider and the horse, for there will frequently be not the slightest warning or opportunity for escape. A horse that has once had the Megrims is very subject to a return of the complaint; for the vessels of the brain, weakened by this violent distention, offer less resistance to the flow of blood, or whatever else may be the cause. No prudent man will drive a horse that has had previous attacks of the disease.

Treatment.—The only thing to be depended on is the use of active cathartics, as directed for inflammation of the brain. After the operation of the physic, medicines of an antispasmodic nature may be given. The following will be very serviceable:

Powdered valerian, . . 1 ounce.
" asafetida, . . 1 "
" ginger, . . 1 drachm.
Water, ... 1 pint.

To be given once each day; or it may be made into a ball by the addition of a little syrup. The food should consist of bran-mashes, green food, or the horse should be turned out for two or three months.

APOPLEXY.

Apoplexy, derived from a Greek word signifying to strike, or knock down, is an abolition of sense and voluntary motion, from suspension of the functions of the cerebrum, &c. The attack sometimes assumes a violent form. The horse falls and dies at once.

Symptoms.—Sometimes the animal falls down suddenly and remains insensible; the eyes are fixed and glassy; breathing becomes laborious; the membranes of the eyes, nose, and mouth are of a dark red or purple color; the pulse is strong, full, and slow; the veins of the neck are distended; the whole muscular system is occasionally affected spasmodically; and the limbs are extremely cold.

The attack sometimes is of a milder nature; the animal will be seen with his head low, extended almost to the ground, and supported against the manger. He staggers as he stands. If moved, he appears to be afraid to fall. His sight and hearing are evidently affected. This is not mad staggers, for no inflammation of the brain is found; nor stomach-staggers, for there is no distention of the stomach. The horse will continue in this way from one to twelve hours; he then falls, grinds his teeth; his eyes are fixed, open and protruded; the pupil is dilated; there are twitchings of the frame; he is unable to swallow; the twitching increases to strong convulsions, and death speedily closes the scene.

Cause.—Compression of the brain appears to be the immediate cause of apoplexy, caused by a congestion of the blood-vessels, or an effusion of serum (water), thrown out between the ventricles of the brain.

Treatment.—Our first business should be to equalize the circulation by the application of the foot-bath, and manage as directed under article *Mode of Producing Perspiration.* The next object should be to evacuate the bowels by means of the following:—

Powdered aloes, . . . 1 ounce.
" mandrake root, . $\frac{1}{2}$ "
" lobelia seed, . . 2 drachms.
Starch water, . . . 1 pint.

to be given as a drench, and repeated in twelve hours if it does not sooner operate. The operation of physic should be assisted by injections of warm water and soap; if this does not succeed, give an injection composed of

Tincture lobelia, . . . 1 ounce.
" capsicum, . . $\frac{1}{2}$ drachm,
Castile soap, . . . 1 "
Warm water, . . 1 quart.

After the operation of the physic no further treatment is necessary; food should be given with great care, and should consist of bran-mashes, carrots, &c.

TETANUS, OR LOCK-JAW.

This is called lock-jaw, because the muscles of the jaw are earliest and most powerfully effected. This is one of the most dreadful and fatal diseases to which the horse is subject. Tetanus is a constant spasm of all the voluntary muscles, and particularly of the neck, the spine, and head. It is generally slow and treacherous in its attack. For a day or two the horse does not appear well; does not feed as usual; partly chews his food and drops it out of his mouth, and gulps his water. The motion of the jaws is considerably limited, and the saliva will drizzle from the mouth.

Symptoms.—The jaws are fixed, and there is a stiffness of the neck. The muscles of the neck are hard and unyielding, and a hardness and prominence can be felt throughout all the muscles, with an unusual protrusion of the head. The nostril is expanded, the ear erect, and the countenance anxious; the back and loins are stiff, and if he turns, the whole body turns at once. The tail is erect and constantly quivering; the extremities are singularly fixed and straddling, like the legs of a stool. The pulse at first is not much affected, but soon becomes quick, small, and irregular; and the breathing becomes more laborious as the disease proceeds, and the countenance is expressive of extreme agony. The disease goes on for eight or ten days, until the animal is exhausted by the expenditure of nervous energy and the continuance of torture.

Cause.—Lock-jaw generally arises from a wound, and oftenest of a wound of a tendon or ligament, but depending not either upon the extent of the wound, or the degree of inflammation. The time of the attack is uncertain, and is frequently postponed until the wound is nearly or quite healed. Tetanus is an affection of the nerves. A small fiber of some nerve has been injured, and the effect of the injury has spread to the origin of the nerve; the brain has become affected, and universal diseased action follows. It occasionally follows nicking, docking, cropping, &c., whether well or ill performed—whether properly attended to afterwards, or neglected. Exposure to cold is a frequent cause; water dropping upon the back through the decayed roof of a stable; or the storm pelting upon the uncovered and shivering animal.

Treatment.—If tetanus arises from nicking, let the incision be made deeper, and the Stimulating Liniment No. 3 be applied; and if it arises from docking, let the operation be repeated higher up. If it be a wound in the foot, let it be touched with the hot iron or the caustic, and keep it open with the Stimulating Liniment. In treating the constitutional disease (that is when it has affected the general system), efforts must be made to tranquilize the system; and the most powerful agent is chloroform, inhaled freely, or until the horse is brought under the full influence of the medicine. It may be applied by means of a sponge; the sponge should be wet with the chloroform and applied to the horse's nose. I never have known this to fail of relaxing the spasm when freely applied. A temporary relaxation of the spasms will always follow, and that will give an opportunity of giving physic, which will serve to quiet the disturbed system, on the principle of counter-irritation and relaxation; and that physic is best which is speediest in its operation. The croton oil has no rival in this respect.

Croton oil, ... 20 drops.
Sweet oil, ... 1 gill.

The medicine should be repeated every eight hours until it operates. In all these nervous affections the bowels are very torpid, and there is little danger of inflammation from drastic physic. The operation of the physic may be assisted by frequent injections. As soon as possible, support the strength by nutritive food. No medicine will be wanted as he gets better; nourishing food, sparingly allowed, will constitute the best tonic; and, if the weather is suitable, the horse should be turned out in the middle of the day.

NASAL GLEET, OR DISCHARGE FROM THE NOSE.

Disease of the mucous membrane of the nose is one of the most common diseases to which the horse is liable. There is a constant secretion of fluid to lubricate and moisten the membrane that lines the cavity of the nose, which, under catarrh or cold, is increased in quantity and altered in appearance and consistence; but that to which we immediately refer is a continued and often profuse discharge, when the symptoms of catarrh and fever have passed away, of a large quantity of thick mucus, sometimes mingled with matter or pus, and either constantly running or snorted out in masses many times in the day.

Treatment.—When the discharge is not offensive to the smell nor mixed with blood, it is merely an increased and somewhat vitiated secretion from the cavity of the nose, and nature will soon cure the disease; but if the discharge is mingled with pus, and offensive to the smell, there is probably ulceration of the mucous membrane, and the horse should be exposed several times a day to the fumes of muriate of ammonia.

Muriate of ammonia, . . . 1 ounce.
Alcohol, ... 1 pint.

This preparation should be dropped on a hot brick, a few drops at a time, and held under the nose of the horse so that he can inhale the fumes. I have cured some of the worst cases of chronic catarrh with this preparation; but if this should not prove successful, there is reason to apprehend that the discharge cannot be controlled, and will soon terminate in

GLANDERS.

The Glanders was described by writers fifteen hundred years ago; it was then, and is now, considered not only a loathsome but an incurable disease. I shall, therefore, be principally confined to the symptoms, nature, and cause.

Symptoms.—The first symptoms of Glanders will be an increased discharge from the nose, of a peculiar nature; it is lighter and clearer, and more glutinous or sticky than in the discharge of catarrh. When rubbed between the fingers, it has, even in its early stage, a peculiar clammy feel; it is not discharged occasionally and in large quantities, like the mucus in catarrh, but is constantly running from the nose. There is a peculiarity about this disease that has never been satisfactorily accounted for; that is, when one nostril alone is attacked, it is in a great majority of cases the near or left. M. Dupuy, the Director of the Veterinary School at Toulouse, gives a most singular account of this. He says, "that out of eight hundred cases of Glanders that came under his notice, only one was affected in the right nostril." The discharge may continue some time before the health and capabilities of the horse seem to be injured. It will remain in a transparent gluey state for some time, and then it will begin to mingle with pus, retaining, however, its sticky character. When pus begins to mingle with the discharge, a new feature of the disease appears. This poison, or virus, is taken up by the absorbents, producing constitutional disturbance. This poison will be absorbed, and the neighboring glands will become affected; and if there be a discharge from both nostrils, the glands within the under jaw will be on both sides enlarged. If the discharge be from one nostril only, the swelled gland will be found on that side alone. Glanders, however, will frequently exist without these swelled glands in its early stage. The glands that are affected in Glanders adhere close to the jaw on the affected side. They are not large and diffused, and in the middle of the channel between the jaw-bones, as in Catarrh and Horse Distemper. The membrane of the nose should be examined to see if there is any ulceration.

How to Distinguish Horse Distemper from Glanders.—In the early stages of Horse Distemper, the horse will appear to have a cold with some degree of fever, and sore throat with some cough and wheezing; and when the enlargement appears between the jaws, it is not a single small gland, but a swelling of the substance between the jaws, growing harder towards the middle; which, after a while, will suppurate and break. After the tumor bursts, the fever will abate, and the horse will speedily get well.

Contagion.—The Glanders is highly contagious, but cannot be communicated by the air or breath. If the division between the horses is sufficiently high to prevent all smelling and snorting at each other, and contact of every kind, and they drink not out of the same pail, a sound horse might live for some years, uninfected, by the side of a glandered one. The matter of glanders has been mixed up in a ball, and given to a healthy horse, without effect. The glanderous matter must come in contact with the mucous membrane of the nose. It is easy, then, accustomed as horses are to smell each other, and drinking out of the same pail, to imagine that the disease may be very readily communicated. Some horses have received the infection from matter blown across a lane, when a glandered horse has neighed or snorted. It is almost impossible for horses to remain long in a stable with an infected horse, without catching the disease.

Treatment.—As for medicine, there is scarcely a drug in the whole materia medica of which a fair trial has not been made, and many of them have had a temporary reputation; but they have passed away, one after the other, and are no longer used. When the disease is local, a solution of white vitriol has done wonders; by syringing it in the nose, it will almost always stop the discharge, and heal up the ulceration. But when the disease has become constitutional, the horse becomes a mass of disease, and should be destroyed.

FARCY.

Farcy is a different modification or development of the same disease; they will run into each other. An animal inoculated with the matter of Farcy will often be afflicted with Glanders, while the matter of Glanders will frequently produce Farcy. They are different types of the same disease. There is, however, a very material difference in their symptoms and progress, and the most important consideration of all, is, that while Glanders is generally incurable, Farcy, in its early stages and mild form, is curable.

Symptoms.—The appearance of small, knotty tumors, termed by farriers, farcy-buds. They appear usually about the face and neck, or inside of the thigh, and in the latter case there is some general enlargement of the limb. When they make their appearance on the inside of the leg, they follow the course of the veins and connect together by a kind of cord which farriers call corded veins; they become hot and painful and begin to ulcerate. These buds have sometimes been confounded with the little tumors or lumps of Surfeit. They are generally higher than those tumors, not so broad, have a more knotty feeling, and are principally found on the inside of the limbs.

Treatment.—In the commencement of the disease give freely of some alterative powder; the following is the most effectual:

  Mandrake root, powdered, . . 4 ounces.
  Blood   "   . . 2   "
  Golden seal "   "   . . 2   "
  Ginger   "   "   . . 3   "

Give a tablespoonful three times a day, in the feed. The horse should be placed under healthy conditions; he should be turned out during the day, or if this is not convenient, the animal should be placed in a large box with a free circulation of air, and should be freely exercised; and green food, and particularly carrots, should be given him, with a fair allowance of oats.

WATER FARCY, OR YELLOW WATER,

Is a dropsical affection of the cellular membrane of the skin, and affects the chest and limbs generally. It is caused by a loss of action in the absorbents throughout the system.

Treatment.—A decoction of the Indian hemp root, is a safe remedy, and very effectual. It will always cure the disease if given in time, and it is a specific for all dropsical affections, whether in man or beast. I have achieved some great cures in dropsy with this medicine, and never fail if called in season.

The following is my way of administering the hemp:

Indian hemp (Apocynum Cannabinum), . 4 ounces.
Boiling water, ... 4 quarts.

After it has stood and got cold, give the horse a quart of the decoction, morning and evening; or you may give one ounce of the fluid extract of hemp, in a pint of water, once a day. This, together with a good nutritious diet, is a sure cure.

INFLAMMATION OF THE EYE.

When the eye becomes inflamed, we should inquire into the cause of the inflammation, the most common of which is high feeding, without a due proportion of exercise. If it arises from any mechanical injury, and this is not very severe, there is a probability of its being speedily removed, by means of the following lotion:

Sugar of lead, ... 1 drachm;
White vitriol, . . . $\frac{1}{2}$ "
Rain-water, ... 1 quart;

to be used as a wash three times a day. But if the inflammation arises from plethora, or redundancy of blood in the system, a sparing diet and depletive remedies will bring the eye all right; but the cure may be assisted by means of the eye-lotion as above directed.

As a depletive remedy, give the Cathartic Powder No. 2 two or three times a week. If this be neglected at the commencement of this disease, though the inflammation, after some time, appears to pass off, and the eye, to superficial observers, seems to have recovered, yet the disease frequently returns, and ultimately occasions blindness.

There is a cartilaginous body connected with the eye of a horse, commonly termed the haw. Whenever the eye is drawn into the socket (which the horse has the power of doing by means of a muscle that does not exist in the human subject), the haw is forced over the eye, so that when dust happens to adhere to the surface of the eye, he is enabled, by means of this cartilage, to wipe it off.

DISEASES OF THE SKIN.

HIDEBOUND.

This term is applied to a tightness of the skin, which feels as if it had grown fast to the ribs, with a rough appearance of the coat at the same time. This complaint is generally caused by worms, or want of condition, and sometimes by a derangement of the digestive organs; and is more properly the effects of disease than disease itself. A few doses of the Cathartic Powder No. 2, together with plenty of oats, will generally set all right again in a short time.

SURFEIT.

This is an absurd term given by farriers to a disease of the skin consisting of small tumors or knobs, which suddenly appear on the surface of the body, sometimes in consequence of drinking largely of cold water when the body is unusually warm, but it appears frequently without any manifest cause. It may be easily cured by wetting the horse's oats in a decoction of sweet fern, alder, and red-topped maple, for a few days. There is another disease of the skin, of the same name, which is more obstinate, and atacks horses that are hidebound and out of condition. In this, a great number of very small scabs may be felt in various parts of the body; the horse is frequently rubbing himself; and sometimes the hair falls off from those parts which he rubs. This complaint approaches to the nature of mange, and requires the same treatment, assisted by regular exercise, good grooming, and a generous diet.

MANGE.

This is not a very common disease, and is seldom met with, except where scarcely any attention is paid to the horse, and when his food is of the worst quality; it is very contagious, and may in this way attack horses in good condition.

Symptoms.—The horse is constantly rubbing or biting himself, so as to remove the hair from the mane and tail. Small scabs are observed about the roots of that which remains. The

Mange is, I believe, a local disease, and requires only local treatment.

Treatment.—Balm of gilead buds, simmered in sweet cream, are a safe and effectual remedy. After applying this ointment two or three times, the horse should be washed all over in castile soap and water. In obstinate cases, however, it may be advisable to try the effects of some alterative medicines; for this purpose give a strong decoction of black alder and sweet fern, or soak the horse's oats with the same.

GREASE.

This disease consists in inflammation and swelling of the heels, and a consequent discharge, having a peculiar offensive smell, and sometimes runs on to ulceration. The swelling frequently extends as far up as the hock and knee-joint.

Treatment.—When the inflammation and swelling are considerable, a poultice should be applied to the heels, and care should be taken that it is kept constantly in contact with the parts. After the inflammation and swelling have abated, the poultice may be discontinued, and the following wash applied several times a day:—

Sugar of lead, ... 4 ounces.
White vitriol, ... 1 "
Crocus martus, ... 1 "
Water, ... 1 quart.

Should the heels be ulcerated, apply the Brown Ointment No. 5. In slight cases of Grease, the lotion and ointment will generally be found sufficient to effect a cure; but when the disease is of long standing, and particularly if the horse has suffered from it before, there will be more difficulty in removing the disease. In such cases, the following powder may be given in the oats every day until it produces considerable alterative effects:—

Mandrake root (powdered), . . 4 ounces.
Cream of tartar, ... 8 "

Dose, a table-spoonful.

Grease is most commonly caused by high feeding and want of exercise, neglect in grooming, &c.; there are cases which seem to depend on general debility. When a horse has suffered much from this disease, and particularly if he appears to be weak and out of condition, a liberal allowance of corn will tend to recover him. In cases of this kind exercise is essentially necessary. Nothing tends so much to prevent Grease and swelling of the legs as frequent hand-rubbing, and cleaning the heels carefully as soon as the horse comes in from exercise. In chronic cases of Grease, where the disease appears to have become habitual, a run at grass is the only remedy.

WOUNDS.

The first operation necessary in wounds is to remove carefully all dirt or other extraneous matter; and if the wound be made with a clean-cutting instrument, and not complicated with bruising or laceration, the divided parts are to be neatly brought together and secured by means of sticking-plasters. Previous to the application of the plaster, the hair should be carefully clipped from off the edge of the wound. After the application of the plaster, the parts should be secured by means of a bandage carefully adjusted, and kept constantly wet with cold water, which is better than all the lotions, balsams, and salves in use. This bandage should not be removed in several days, that the divided parts may have time to heal by the first intention, as surgeons term it, unless considerable inflammation and swelling come on; it would then be advisable to remove the bandages and apply fomentations; but this is seldom necessary where cold water has been freely applied at first. This kind of union, however, can seldom be accomplished in the horse, from the difficulty of keeping the wounded parts in contact, and from their wounds being generally accompanied with contusion or laceration; yet it should always be attempted. When the swelling and inflammation are removed, the fomentations and poultice are no longer necessary, and the Horse Liniment No. 5 should be applied. Should the wound not be disposed to heal, discharging a thin and offensive matter, apply the Brown Ointment No. 5. If the granulations become too luxuriant (that is, if what is commonly termed proud flesh makes its appearance), burnt alum is to be sprinkled on the wound.

Slight wounds will heal of themselves, or without the interference of art; and it is from this circumstance that many nostrums have acquired unmerited reputation.

If a blood-vessel of any size is wounded, and the bleeding is likely to prove troublesome, our first object should be to stop the bleeding, which, if the wound be in a situation that will admit of the application of a roller or bandage, may be easily effected; for pressure, properly applied, is generally the best remedy on these occasions, and far more effectual than the most celebrated styptics. A puff-ball bound on by means of a bandage is very effectual. I have succeeded in stopping blood, even when bleeding from an artery, by syringing in ice-water, and never knew it fail if thoroughly applied. In some cases it becomes necessary to tie up the bleeding vessel; this is rather a difficult operation, and requires the assistance of a surgeon.

Punctured Wounds, or such as are made with pointed instruments, are generally productive of more inflammation than those that have at first a more formidable appearance; and if such wounds happen to penetrate a joint, or the cavity of the chest, or belly, the worst consequences are to be apprehended.

When the synovia, or joint-oil, is observed to flow from the wound, we may know that the ligament of the joint has been punctured. The first thing to be done in this case is to close the opening that has been made into the joint; for as long as it remains open the inflammation will increase, and the pain will be so violent as to produce a symptomatic fever, which often proves fatal. The most effectual method is to syringe in the wound balsam of copaiba; this is perfectly safe. No stimulating or caustic preparations should be used. The old farriers often injected caustics into the joint, and thereby brought on fatal inflammation and most excruciating torments. Sometimes their patients were destroyed by fever, which followed; more frequently, however, the joint became stiff or immovable, and the wound healed.

Punctured wounds of the feet are most frequent, and are caused, either by the horse stepping on a nail, or *picking up* a nail, as it is termed, or by carelessness in shoeing. In the former case the nail generally enters the frog, and often penetrates the joint of the coffin-bone. The sole is generally sufficiently hard to resist the nail, but the frog is commonly of a softer and more spongy nature. When the coffin-joint is wounded, there is danger of an incurable lameness from the joint becoming stiff; but by proper management the wound is often closed in a short time, and the free use of the joint preserved. When the foot is wounded by a nail we should immediately open the orifice in the sole. If the joint be wounded, synovia, or joint-oil, will issue from the wound, but in small quantity. After the orifice is made sufficiently large, the balsam of copaiba should be applied, and bathed well in by means of a hot brick. But when the joint has escaped injury, after enlarging the opening made by the nail in the sole, and cutting away the horn from the contagious parts until it becomes very thin, a little tow or lint dipped in tar should then be applied, and the whole foot kept cool by means of a bran poultice, or kept constantly im-mersed in lukewarm water. The most essential part of the treatment consists in opening well the orifice into the horny matter; for in wounds of this kind we always find that soon after the nail has been withdrawn the puncture in the horn closes or nearly closes up, but the living parts that have been wounded under the sole soon inflame and swell; consequently they suffer considerable pressure, as the horn is too thick and inflexible to give room for them as they swell. When matter forms, it being confined by the horny envelope, diffuses itself between the sensible and insensible parts, so extensively some-times as to cause a separation of the entire hoof. When it has been found necessary to remove some part of the horny sole, in consequence of matter having been formed under it, a pledge of tow dipped in tar and lard should be applied.

The farrier sometimes wounds the sensible part of the foot in shoeing. The horse is then said to be pricked. The nail, instead of being driven into the horny insensible part only, is either forced into the living parts, or so near to them that its pressure gives such pain to the animal as to cause him to go lame; inflammation gradually takes place in consequence, and at length matter forms, which, if not allowed to escape, by remov-ing the shoe and cutting away the horn, spreads under the hoof, and after some days breaks out at the coronet. When the horse has been pricked, the mischief is not always discovered imme-diately after shoeing. The pressure upon the insensible part is sometimes too still to cause lameness at first; so, when the horse is observed to go lame, the farrier pronounces it to be in the shoulder, and the poor animal is tormented by the strong liniments or blisters applied to the shoulder, while he is suffer-ing from another cause. Consequently the disease is allowed to run to such lengths as we have described.

Punctured wounds in other parts of the body are often inflicted with the stable-fork, either accidentally or intentionally. I have often known joints wounded in this way. When the flesh only is punctured, the orifice must be kept open, that the wound may heal from the bottom; and if the sides become callous and not disposed to heal, a mild stimulant may be used with a syringe, such as tincture of blood-root, or tincture of benzoin. In recent wounds, however, of the punctured kind, these irritating applications are improper; wounds of this kind are frequently followed by considerable pain and inflammation. In view of this we should keep the orifice open, and if it be small, to enlarge it with a lancet, and keep it open until the pain and inflammation have subsided.

BRUISES.

In recent bruises, before much swelling or inflammation takes place, cold applications are undoubtedly the best, and, if thoroughly applied, will generally prevent all swelling and inflammation. But, after much swelling has taken place, fomentations are the most proper remedies. When bruises are violent, a considerable degree of inflammation may be expected to supervene. It will then be proper to give physic and take off the horse's feed.

If abscesses form in consequence of a bruise, discharging large quantities of matter, and particularly if the matter be of a bad color and an offensive smell, the wound also appearing dark-colored and rotten, indicating approaching mortification, the animal's strength must be supported by allowing him a plenty of oats, carrots, &c. It will be necessary also, to give tonics and stimulants. The following will admirably fulfill the indication:

Golden seal, powdered . . 4 ounces
Ginger root " . . 2 "
Blood " " . . 1 "

Give a table-spoonful three times a day. Stimulating applications to the parts, such as aqua ammonia or camphorated spirits, are of great use.

If after the inflammation has subsided, a hard, callous or swelling remain, apply the Restorative Balsam, No. 3, and bathe it in well three times a day.

FISTULA IN THE WITHERS.

This affection is generally caused by bruises from the saddle, and is at first a simple abscess, which by proper treatment and early attention may be easily cured; but if neglected it runs into a fistulous sore, proves extremely difficult to cure, and requires severe treatment.

As soon as the swelling is discovered, fomentations should be applied in order to promote suppuration; and when matter is formed, let the tumor be opened so that its contents may escape, and a future accumulation prevented.

The sore may now be healed by dressing it daily with the Brown Ointment, No. 5. I have generally found this ointment effectual. But when the disease has been neglected in its first stage, and the ligaments and bones of the withers have become affected, a more severe treatment is required. In this case inject tincture of iodine into the cavity several times, and keep the head constantly wet with cold water by applying a pad on the top of the head, and securing it by means of a bandage. The sore is not to be dressed until the sloughs which the tincture occasions have separated from the living parts, which generally happens two or three days after the application of the iodine. If the surface of the sore looks red and healthy, and the matter appears healthy and looks white and of a better consistence, a repetition of the iodine is not required, and the Brown Ointment will be sufficient to complete a cure; but if it should still retain an unhealthy appearance, and the matter continue thin and of a bad color, the iodine tincture must again be applied

POLL-EVIL.

When we consider the weight and position of the horse's head, with the great length of the neck, it will readily appear that the muscles alone are not capable of supporting and moving so great a weight, under such mechanical disadvantages. Nature has therefore provided a strong ligament which is firmly attached to the back part of the head, whence it passes down over the bones of the neck. It is not attached to the first bone, but is firmly fastened to the three next; it then passes over the three other bones of the neck in nearly a straight line to the withers, where it is securely fixed by giving off a thin slip of ligament in its passage, which is united to the last three bones. It is continued from the withers to the back. This ligament, being elastic, allows of sufficient motion in the neck, and so effectually assists the muscles in supporting the head, that they never become fatigued.

A poll-evil generally originates from a violent blow on that part of the poll which covers the first bone of the neck, which, as we have just observed, is not attached to the ligament, the injury will be chiefly sustained by the sensible parts placed between the bone and the under surface of the ligament. The skin may also be hurt, and a slight degree of superficial inflammation take place. But when inflammation has been thus produced between the bone and the ligament, it is more likely to proceed to suppuration, or to the formation of matter, which, being deeply seated, cannot find vent at the surface, like a common abscess; therefore it spreads under the ligament, and is so long in arriving at the surface, that both the bones and ligaments are highly diseased before any external swelling is observed. I am convinced, from observation, that it is almost impossible to disperse the genuine poll-evil.

Treatment.—In attempting to disperse a poll-evil we lose much valuable time, and suffer the matter to continue its ravages upon the ligaments and bones. The only effectual practice consists in opening the abscess freely, so that the matter may readily escape, and the parts be examined. When this has been done, and the bleeding has perfectly ceased, apply the Brown Ointment No. 5, and let the first dressing remain until the dead parts are ready to separate. It is sometimes necessary to repeat the dressing several times. When the wound has been brought to a healthy state it will heal of itself, or nature is sufficient to effect a cure.

Horses are frequently calloused on the top of the head, caused by tight checking. This constant stretch on the poll, or ligament of the neck causes a slight degree of inflammation, which, instead of going on to separation, terminates in scirrhus or hardening. When we are called upon to treat poll-evil after it has broken out, and a pipe is formed, put white vitriol in a goose-quill, and introduce it into the pipe. The quill should be cut off at the little end, so that the vitriol may act on the lower part of the pipe. When the pipe is eaten off at the bottom, it will shove up. During this operation, cloths should be kept constantly wet with cold water on the horse's head, to keep down inflammation. After the pipe has been destroyed, use the Brown Ointment No. 5, and it will soon heal.

STRAINS.

Every person ought to be well acquainted with this subject, since his horse is particularly liable to such accidents. Strains may affect either muscles, ligaments, or tendons. Muscular strains consist in an inflammation of the muscles or flesh, occasioned by sudden and violent exertion. When a ligament is strained, there is generally some part of it ruptured, causing obstinate and sometimes permanent injury. In this case, also, inflammation is the symptom which requires our attention. But tendons are the parts most frequently affected, and particularly the flexors of the fore-legs, or back-sinews, as they are commonly termed. Tendinous strains consist in an inflammation of the membranes in which tendons are enveloped, and the swelling which takes place in these cases depends on an effusion of coagulable lymph by the vessels of the inflamed part. Inflammation being the essence of a strain, we are to employ such remedies as are best calculated to subdue it; and should any swelling remain, it is to be removed by stimulating the absorbent vessels to increased action.

STRAIN OF THE SHOULDER.

A strain in this part is by no means so frequent as is generally supposed, lameness in the feet being often mistaken for it. The symptoms are so well marked that a judicious observer will never be at a loss to distinguish one from the other. A strain of the shoulder is an affection of some of the muscles of the shoulder—most commonly, I believe, those by which the the limb is connected with the body. When the shoulder is strained, the lameness comes on rather suddenly, and it is generally considerable. When the horse attempts to walk, the toe of the affected side is generally drawn along on the ground, from the pain which an extension of the muscles occasions. In violent cases, he appears to be incapable of extending it. When the lameness is in the foot, it is generally gradual in its attack, unless occasioned by an accident, and does not at all hinder the extension of the limb; an unusual heat and tenderness may also be perceived in the foot, and as the horse stands in the stable the affected foot will be put forward, that it may bear as little weight as possible.

The first remedy to be employed in this case is the Horse Liniment No. 5. This should be applied thoroughly, and well bathed in with a hot brick. If the injury be considerable, a rowel should be put in the chest. By these remedies, and rest, the lameness will generally be removed in a short time. A cooling, opening diet, with perfect rest, will be necessary. When the inflammation and lameness begin to abate, the horse should be turned into a loose stall, and after a week or two he may be suffered to walk out for a short distance every day. The use of moderate exercise after the inflammation is in a great measure subdued, is to effect an absorption of any lymph that may have been effused, and to bring the injured muscles gradually into action.

STRAIN OF THE STIFLE.

The stifle-joint is strained more frequently than any other part of the horse; the location of the injury may be known by the stifle-joint being hot and tender, and sometimes swollen, and the horse will not place the foot on the affected side as far forward when he moves as he does the other one.

The remedies are, fomentations and astringents. The following preparation will be found very beneficial:

Tincture of arnica, ... 4 ounces.
Alum, ... 1 "
Water, ... 4 "

To be applied frequently. In a chronic affection of this joint, where the tendons and muscles have become relaxed, a plaster composed of

Powdered alum, ... 2 ounces,
Common salt, ... 1 "

with the whites of two eggs, mixed up with flour to the consistency of a thick paste, and rubbed well in over the stifle-joint, will be of great service; or a common strengthening-plaster may be applied, and let it remain until it comes off itself. Strain of the hock-joint requires the same treatment.

STRAIN OF THE HIP OR WHIRL-BONE JOINT.

Lameness of the hind leg is too obscure for the farrier's comprehension; consequently he generally pronounces it to be a strain in the round or whirl-bone joint, with all that affectation of infallibility so commonly observed in those gentlemen. I have seen several cases of lameness which were supposed to be occasioned by an injury of this part, but after attentive examination it proved to be a spavin.

The hock-joint in such cases should be carefully examined, and if unusual heat or tenderness be observed on the seat of the spavin, it is probable that the lameness arises from this cause. I have met with several cases where the horse had been severely burnt and blistered in the hip, when the hock was evidently the seat of the lameness.

Treatment.—Rest and stimulating liniments are the most proper treatment. The Stimulating Liniment No. 3 will be found a valuable application, and should be used for some length of time. There is no lameness that requires perfect rest as much as lameness of this joint.

STRAIN OF THE FLEXOR TENDON, OR BACK SINEW.

This is one of the most common of all strains, owing to the fashionable construction of the stable and the common custom of shoeing, with the toes left long and heels paired down low, and the stable-floor raised six or eight inches the highest before. Just imagine the condition of the horse kept constantly in the stable, and the constant stretch of the back sinews thus conditioned, together with the violent or long-continued exertion required of the horse, and it will not be wondered at that they frequently give way. Some of the fibers which tie the tendons down are ruptured. A slight injury of this nature is called a strain of the back sinews or tendons, and when it is more severe the horse is said to have broken down.

The first injury is confined to inflammation of the sheath or membrane, and a rupture of a few of the attaching fibers. The inflammation of the part, however, is often very great, the pain intense, and the lameness excessive. The anguish expressed at every bend of the limb, and the local swelling and heat, will clearly indicate the seat of the injury.

Treatment.—The first object in view should be to abate the inflammation of the part, and no means are so likely to succeed in the first stages as bandages kept constantly wet with cold water; but if the difficulty has been of several days' standing, and considerable swelling has taken place, the parts should be well fomented with warm water several times a day, or it would be better to keep it constantly applied. After the fomentations are discontinued, the legs should be inclosed in a poultice of linseed-meal. After the horse begins to put his foot to the ground, and to bear pressure on the part, the enlargement must be got rid of, and the parts must be strengthened. The two latter purposes cannot be better effected than by using an elastic bandage; one of thin flannel will be the best. The Stimulating Liniment No. 3 should be applied previous to the application of the bandage. By these means the absorbents are sooner induced to take up the effused coagulable matter of which the swelling is composed. The bandage should be daily tightened in proportion as the parts are capable of bearing increased pressure, and the treatment should be persisted in until the swelling disappears; after this has been effected, the horse may gradually be put to his work.

RUPTURE OF THE SUSPENSORY LIGAMENT.

By extraordinary exertion, the suspensory ligament is sometimes ruptured. In this case the fetlock almost touches the ground, and it is sometimes mistaken for rupture of the flexor tendon; but one circumstance will sufficiently demonstrate that it is the suspensory ligament which is concerned, viz., that the horse is able to bend his foot. Rupture of this ligament is a bad, and almost desperate case. The horse is frequently lame for life, and never becomes perfectly sound.

Treatment.—The horse should be kept perfectly quiet, and the leg should be well bandaged, and a high-heeled shoe put on, which will afford temporary relief.

RING BONE

Is a bony excrescence about the small pastern bone, near the coronet, or an ossification of the cartilage of the foot. When it is observed in its early state, it may be cured by the following receipt:

Spirits of turpentine,	8 ounces,
Spanish flies,	$\frac{1}{2}$ "
Gum camphor,	$\frac{1}{4}$ "
Sal ammoniac,	$\frac{1}{4}$ "

This is by no means an infallible remedy; the complaint is

frequently incurable, and especially when the disease has proceeded so far as to cause a stiff joint, there is no chance of cure.

THOROUGH-PIN.

A thorough-pin is a fullness on both the inside and outside of the hock-joint. When we press on one side of the tumor, the fluid which it contains is forced into the opposite. From this circumstance the disease has probably taken its name.

Cause.—It is generally caused by hard work, and, therefore, difficult to cure.

Treatment.—The best remedies are rest and the following preparation:

Spirits of turpentine,	8 ounces,
Spanish flies,	¼ "
Gum camphor,	½ "
Sal ammoniac,	" "

WINDGALL

Is a vuglar name given to an enlargement of the mucus sacs, which are placed behind the flex or tendons, for the purpose of facilitating their motion; and the ancients supposed that they contained air or wind. If punctured, they discharge a fluid resembling joint-oil; and they frequently communicate with the cavity of the joint, and, therefore, cannot be opened without danger of producing incurable lameness.

Treatment.—Bandages applied to the parts, kept constantly wet with cold water, have a good effect; or use this embrocation:—

Sugar of lead,	4 ounces,
White vitriol,	2 "
Alum,	1 "
Water,	1 quart.

This complaint does not often occasion lameness, and is, therefore, seldom much attended to; but as it is almost always a consequence of hard work, and often renders a horse unfit for much labor, it diminishes his value considerably.

SPLINTS.

Splints consist in a bony tumor about the shank-bone, between the knee and the fetlock-joint; they seldom occasion lameness unless situated so near the knee or back-sinews as to interfere with their motion.

There are many cases of lameness that are atributed to splints, when the cause evidently exists in the foot.

The bony tumor may sometimes be removed by the application of the Spavin Ointment No. 4. This should be used once a day for some time.

SPAVIN.

A spavin is an enlargement on the inside of the hock, and is of two kinds :—The first is a bone-spavin, consisting of a bony tumor ; the other, a bog or blood spavin, and is an enlargement of the mucus sacs which communicate with the joint. The former often occasions lameness just before it makes its appearance, and then can be discovered only by feeling the part, which will be found unusually hot and tender.

Treatment.—If the Spavin Ointment No. 4 be applied at this period of the disease, and bathed well once a day, it will generally prove successful.

The bog-spavin does not often cause lameness, except when the horse is worked hard ; this causes a temporary lameness, removed by rest, but it does not often admit of a radical cure; for though it is frequently removed by medicine, it generally returns when the horse is made to perform any considerable exertion.

CURB.

This is a bony tumor on the back part of the hock, which sometimes occasions lameness.

Treatment.—All stimulating liniments and blisters should be avoided in the first stages of this disease, and cooling washes applied, such as

Sugar of lead, ...	4 ounces,
White vitriol, . .	1 "
Water, ...	1 quart.

After the inflammatory stage has passed by, and an enlargement remains, treat it as you would a bone-spavin.

DISEASES OF THE FOOT.

Of all diseases to which the horse is liable there are none that occur so frequently, and none more difficult of cure, than those which attack the foot; and however strange it may appear to those who have not paid much attention to this subject, it is an incontrovertible fact that almost all of them are the consequence of bad shoeing, and improper management of the foot. It is one of the most difficult things in the world to find a good horse-shoer; they are almost without an exception perfectly ignorant of the anatomy and physiology of the horse's foot, or the use of the various parts that compose it. It is necessary to be well acquainted with the natural form of the foot, in order to determine how far it has been altered or destroyed by disease or bad shoeing. The most frequent cause of lameness in in the foot is a contraction of the horny matter that composes the hoof, generally accompanied by an increased concavity and thickness of the sole. The cavity of the hoof being thus diminished, the sensible foot suffers a greater or less degree of compression, which occasions in it inflammation and lameness.

When we examine the bottom of a contracted foot, instead of being circular it will be found of an oblong form; the heels and frog will appear as if they had been squeezed together. In some cases the frog has become rotten and discharged matter.

We sometimes meet with horses that go perfectly sound, though their hoofs are much contracted; on the other hand we often see severe lameness produced by a slight degree of contraction.

Treatment.—In attempting to cure this disease we should first remove carefully with a knife all the rotten parts of the frog, and apply tar and tallow to those which are sound. A small quantity should be poured into the cleft of the frog; this will promote the secretion of horny matter, and, if assisted by pressure, will increase the solidity of that already formed. The quarters and heels should then be rasped, particularly at the coronet, and kept constantly wet with cold water; and expose the frog constantly to pressure, either by means of the artificial frog, or by reducing the crust at the heels. In treating contracted feet, nothing is so important as a run at grass, which places the horse under natural conditions; soft wet ground should be selected. Previous to being turned out the horse should have his shoes removed, so that the bottom of the foot can come in contact with the ground, taking care that the bottom of the foot is occasionally reduced, so that the frog may constantly receive pressure. When the horse runs at grass, the hoof keeps constantly wet and soft, which is better than all the hoof ointments in existence. Ointment should not be applied, for the reason that all oily preparations prevent the action of the water on the hoof; consequently you lose the benefit in a great degree of a run at grass. Water is the only diluent in nature, and the only thing that will keep the hoof soft and natural. A hoof cannot contract as long as it is kept wet; neither can a board, but when kept dry it will contract and warp. A horse that is kept up in the summer should always be kept on the ground, so it can be kept natural and cool,

The next disease to be noticed is flat and convexed soles, or, as it is more commonly termed, the pumiced foot. Heavy draft-horses are most subject to this disease, which seems to arise from a weakness of the crust; for when the sole becomes flat or convexed, the crust also loses its proper form and becomes flatter, appearing as if it had been incapable of supporting the animal's weight, and had therefore given way, allowing the internal foot to press so upon the sole as to give it the appearance we observe. The cause of the disease will be better illustrated when we consider that when a horse is drawing a heavy load, not only his own weight, but a great part of that which he is drawing, also, is thrown upon his feet; and as the fore-feet support by far the greatest share, it is not at all surprising that the crust should sometimes give way. Although it possesses sufficient strength for the purposes of the animal in a state of nature, yet the strength is limited, and not always capable o fsustaining itself under the heavy burdens which the crust has to sustain. The sole is rendered thinner, when it becomes flat or convexed, than it naturally is, and sometimes so much so as to yield to the pressure of the fingers.

Treatment.—In attempting to cure this disease, it is first necessary to take off the horse's shoes, and make him stand on a flat, hard surface; this kind of pressure will harden the sole, and in the end render it thicker, particularly if tar or turpentine be frequently applied to them. We frequently meet with horses whose pasterns are remarkably long and slanting in their position, while their heels are very low, and the toe of considerable length. If thin-heeled shoes were applied to feet of this description, or if the toes were not kept short, the horse would be very liable to lameness, from the extraordinary pressure to which the ligaments and back sinews would be exposed; the heels, therefore, of such horses are to be carefully preserved, and the toes kept as short as possible. The shoes which are applied to such feet should be made sufficiently thick and long at the heel to make up for the deficiency of the horn at that part, in order to relieve the ligament and back sinews; and with the same view the toe should be made rather thin, and of the best steel, and very flaring or concaved.

We frequently meet with another kind of deformity in the foot; that is, the hoof loses the oblique form of a natural foot, and approaches towards the perpendicular form; at the same time the heels become very high. In this case it is necessary to reduce the crust at the heel, and apply thin-heeled shoes

CORNS

Corns are not a disease, but the effect of bad shoeing or improper management of the foot, and may be avoided by following the directions I have given under that head.

Treatment.—When they do occur, it is necessary to remove the cause that produced them, and the effect will generally cease by placing the horse under natural conditions; but sometimes we are under the necessity of removing the red parts, or corns, with a knife, and applying the shoe so that the tender part may not receive any pressure. When this has been neglected, we sometimes find matter formed in this part, which often breaks out at the coronet. In this case it is necessary to make an opening for the matter in the angle between the bar and the crust. Corns may be cured by the application of spirits of salt, as a common thing, without the use of the knife.

QUITTER.

Quitter generally arises from a wound or bruise in the coronet, and, when neglected, penetrates under the hoof, forming sinuses or pipes in various directions.

Treatment.—The most effectual method of treating this complaint is to force into these pipes, by means of a syringe, spirits of salt or muriatic acid. This, though apparently a severe remedy, will be found very effectual. Any caustic application would effect a cure, but I have succeeded so well with the spirits of salt that I have not been induced to try any other medicine. It sometimes requires several applications of the medicine, but generally one application is sufficient to destroy the pipes. When this object is accomplished, nature will heal the sore.

THRUSH.

This is seldom, if ever, an original disease, but merely a symptom or an effect. The cause is generally a contraction of the horny matter at the quarters or heels, by which the sensible frog is compressed and inflamed.

Symptoms.—A discharge of fetid matter from the cleft of the frog, which part is generally rotten, and so soft as to be incapable of affording sufficient protection to the sensible frog, which it covers; hence arises that tenderness of the foot. The inflammation and discharge which take place are the efforts of nature to cure the disease. The discharge should not be too hastily stopped by the application of powerful astringents, which would cause inflammation and severe lameness, which are generally removed by a return of the discharge.

We cannot treat this disease successfully without first removing the cause that produced it. With this view, the quarters should be rasped, and the hoofs kept constantly wet, taking care to keep the frog dry by means of tar. When in this way we have succeeded in removing the compression and consequent inflammation of the frog, it will be advisable to apply the following lotion to the frog:—

Sugar of lead, ... 4 ounces.
White vitriol, ... 1 "
Common salt, ... 2 "

But it will be of no use as long as the cause remains that produced the disease, or until you place the horse's foot under healthy conditions.

FOUNDER.

The foot, like every other part, is liable to inflammation from various causes, and particularly from violence, long-continued action, and more especially letting the horse drink cold water when very warm from exercise.

Treatment.—As soon as the horse is discovered to be foundered, he should be placed in a vat of water about summer temperature; this is better than to place the horse in running water, as this chills the horse and obstructs the circulation.

When it gets too warm from the heat of the feet, it should be changed; it should be kept two or three degrees below the temperature of the body, and the horse should be kept constantly in it until all fever and inflammation have abated. If the horse cannot be kept on his feet, they should be kept constantly wet by means of cloths or a sponge. This treatment is an infallible cure, and will be found more effectual than all the medicines in the whole materia medica. It requires no medicine, excepting a dose of physic occasionally

OF PHYSIC.

In the administering of this class of medicines, great care and attention are necessary, the bowels of the horse being particularly irritable and liable to inflammation. The physic commonly given is certainly too strong, and many horses have been destroyed by the large quantities that have been recommended by writers on farriery. A modern author has ingeniously availed himself of this prejudice to explain the violent effects which his cathartic prescriptions had produced. I am convinced, from experience and observation, that those effects were occasioned by the excessive quantity rather than by the impurity of the cathartic medicines.

The most effectual and safe purgative for the horse is the Catharic Powder No. 10, and it will apply in more cases than any catharic combination now in use.

It is advisable, when time can be spared, to prepare a horse for physic by giving him bran-mashes for a day or two. This will gently relax the bowels, and remove any hardened feces that may be lodged in them; it will also tend to facilitate the operation of the medicine.

When a horse is purged for the first time, he should have a moderate dose; were the common quantity given to one of weak and irritable bowels, there would be danger not only of producing great debility, and thereby counteracting the intention of the medicine, but likewise of destroying the animal, by bringing on an inflammation of the bowels; and this is by no means an unusual occurrence. Although the stomach of the horse and first part of the intestinal canal are not so susceptible as in man, yet the latter part of the bowels is far more so, and more liable to be irritated and inflamed from the effects of drastic catharics.

The morning is the best time to give a purgative, the horse having previously fasted two or three hours. If he be disposed to drink after taking a ball or catharic powder, give a moderate quantity of warm water, which will promote its solution in the stomach, and consequently expedite the operation. During this day the horse is to be kept in the stable, and fed with bran mashes, and a moderate quantity of hay; he should be allowed to drink plentifully of warm water, but if he refuses it give him cold water. The following morning he is to be exercised, and at this time the medicine will generally begin to operate. Should the purging appear to be sufficient, he need not be taken out the second time; but when the desired effect does not readily take place, trotting exercise will tend to promote it. When physic does not operate at the usual time, the horse appearing sick and griped, relief may generally be obtained by giving a clyster of warm water or thin gruel, and let him drink freely of warm water, and have exercise. If the physic continues to operate longer than desired and the horse appears to be considerably weakened, four ounces of paregoric should be given. As a matter of convenience I have given below some good and effectual combinations of physic.

PHYSIC BALLS.

Powdered aloes, . . . 1 ounce.
" mandrake, . . $\frac{1}{4}$ "
" ginger, . . . $\frac{1}{4}$ "

Honey or mucilage sufficient to form the whole mass into balls. To be given at one dose.

Another.

Powdered aloes, . . . 1 ounce.
" blood root, . . $\frac{1}{2}$ "
" ginger, . . . $\frac{1}{2}$ "
Oil peppermint, . . 5 drops.

This composition may be formed into a ball by the addition of syrup or honey.

Another.

Powdered aloes, . . . 6 drachms.
" mandrake, . 2 "
Oil peppermint, . . . 5 drops.

ALTERATIVES.

Alteratives are a class of medicines which produce their effects almost insensibly, and do not interfere with the food or work of the horse. The following formula will be found very efficacious:—

Powdered blood root,	6 ounces.
" dandelion root,	8 "
" mandrake "	4 "

Mix for eight doses.

Another.

Powdered mandrake root,	6 ounces.
Cream of tartar,	8 "

Mix for eight doses.

Another.

Powdered aloes,	8 ounces.
" blood root,	4 "
" lobelia seed,	2 "
" capsicum,	1 "

Mix for ten doses.

Another.

Powdered mandrake,	2 drachms.
" blood root,	3 "
" blue flag root,	1 "

To be given in bran or oats, at a dose, three times a day.

Should a ball be thought more convenient than a powder, the change may be easily made by the addition of syrup or even flour—flax seed meal is the best.

A dose of the alterative powder should be given every evening with oats or bran until the whole quantity is used.

Another.

Fluid extract of dandelion,	1 ounce.
" " sarsaparilla,	1 "
" " blue flag,	2 drachms.

To be given in a pint of water as a drench.

LAXATIVES.

By the term laxative is meant an opening medicine that operates very mildly, and produces so gentle a stimulus upon the intestines, as merely to hasten the expulsion of their contents without increasing the secretions. Linseed oil is the best medicine of this kind, though olive oil will produce nearly the same effect. A dose of the former is about one pint, but the latter may be given to one quart.

When a laxative ball is required, the following will be found useful:—

Powdered aloes, . . 4 drachms.
Castile soap, . . . 2 "

Syrup enough to form a ball for one dose.

DIURETICS.

Diuretics are a class of medicines that stimulate the kidneys to an increased secretion of urine. The following formula I have found convenient and efficacious:

Balsam copaiba, . . . 1 ounce.
Spirits nitre, . . 2 "
Oil almonds, . . . 2 "
Spirits turpentine, . . 1 "
" camphor, . . . 1 scruple.

Give a tablespoonful of the above, in a pint of starch or flax-seed tea, once a day.

Another.

Queen of the meadow, . . 2 ounces.
Milk weed, . . . 2 "
Dwarf elder, . . . 3 "

Bruise, and make a strong decoction by adding one quart of boiling water; when cool, give it at one dose.

SUDORIFICS.

These are medicines that act on the capillary vessels of the skin and cause perspiration. The following formula is one of the most convenient and effectual that I have ever made use of:

Gum opium, . . . 1 drachm.
" camphor, . . 2 "
Ipecac, . . . 1 "
Cream of tartar, . . 1 ounce.

Pulverize all fine, mix, and give one table-spoonful at a time.

TONICS.

These are medicines that increase the strength or tone of the animal system; obviating the effects of debility, and restoring healthy functions

The following formulas can be relied upon as very convenient and efficacious:

  Golden seal root, . . . 4 ounces.
  Unicorn " . . . 2 "
  Ginger " . . . 1 "

Powder fine, and give one ounce three times a day.

Another.

  Powdered columbo root, . 6 ounces.
   " gentian " . . 2 "

Mix, and give in half-ounce doses.

Another.

  Blood-root, powdered, . . 1 ounce.
  Peruvian bark " . 2 "
  Ginger " . . . 1 "

Mix, and give in one-ounce doses.

STIMULANTS

Are medicines which produce an increased and quickly diffused vital energy, and strength of action in the heart and arterial system. The following composition is the best combination that can be given:

  Bayberry bark, . . . 6 lbs.
  Ginger root, ... 3 "
  Cloves, ... 6 ounces.
  Capsicum, . . . 5 "
  Cinnamon, ... 6 "

Pulverize fine, mix, and give in one-ounce doses.

ANTISPASMODICS

Are remedies for spasms or convulsions. The following preparation is very efficacious:—

  Tincture of asafetida . . 1 ounce.
  " " valerian . . 2 "
  " " ginger, . . 1 "

Another.

The inhalation of chloroform is one of the most effectual of all known medicines, yet should be administered with caution. It is a sure remedy in tetanus or lock-jaw, which is one of the worst spasmodic actions known.

RUBEFACIENTS.

These are medicines which, when applied to the skin of a horse, produce increased action in the vessels of the skin, without blistering; such as alcohol, stimulating liniments of various kinds, tincture of capsicum, turpentine, &c.

They are indicated in all internal inflammations and congestions, in view of counter-irritation; for lameness of long standing, and to soften hardened tumors, &c.

VESICANTS, OR BLISTERS.

These are preparations which produce a violent action on the skin, and cause the capillary vessels to throw out an increased quantity of fluid.

BLISTERING OINTMENTS.

No. 1.

Spanish flies, powdered,	½ ounce.
Oil of turpentine,	1 "
Ointment of wax or lard,	4 "

Mix.

No. 2.

Oil of turpentine,	2 ounces.

To which add gradually,

Vitriolic acid,	2 drachms.
Spanish flies, powdered,	1 ounce.
Hogs' lard,	4 "

Mix.

No. 3.

Hogs' lard,	6 ounces.
Venice turpentine,	4 "
Beeswax,	2 "
Yellow rosin,	1 "
Oil origanum,	1 "
Powdered cantharides,	3 "

Melt the first three ingredients; and when removed from the fire, and partly cool, stir in the origanum, and cantharides; continue stirring until cold. Should this blister become too hard in winter, it may be softened by rubbing it with oil of origanum or a little turpentine.

STYPTICS

Are medicines that have an astringent quality which produces contraction and stops bleeding. Among the most valuable are, gum benzoin, tannin, alum, sugar of lead, oil of whipsewog or milk-weed; the latter is the most effectual of all styptics for internal or external hemorrhage.

SEDATIVES

Are medicines which allay irritability and irritation and irritative activity, and which assuage pain and decrease arterial action. One of the most active and powerful is veratrum; the next active is digitalis, or fox-glove. These medicines should be given in all high inflammatory action.

DISCUTIENTS

Are medicines or applications which disperse a tumor or any coagulated fluid in the body. The following has been used with great success.

Mutton tallow,	.	6 ounces.
Bayberry wax,	.	4 "
Iodine,	.	½ "
Hydriodate of potassa,	.	1 "

Melt the tallow and wax; and when cool, rub the whole together. This preparation is excellent for enlarged glands of all kinds.

ANTISEPTICS

Are medicines which resist or correct putrefaction, or correct a putrescent tendency in the system; among which are cinchona, acids, and saline substances, such as salt, chloride of lime, and soda.

FOMENTATIONS.

Fomentation is the application of warm liquors to a part

of the body, by means of flannels dipped in hot water or medicated decoctions, for the purpose of easing pain, by relaxing the skin, or of discussing tumors. They should be applied as hot as can be, without giving pain to the animal.

POULTICE.

A Poultice is soft composition of meal, bran, or the like substances, to be applied to sores, inflamed parts of the body, &c. They are commonly made by boiling wormwood, southernwood, chamomile flowers, and smart-weed, in water, straining off the water, and thickening with bran or linseed flour.

ROWELS.

Roweling is the act of piercing the skin, and inserting a roll of hair or silk. When these are used with a view of relieving internal inflammation or fever, it will be found useful to apply blistering ointment, instead of turpentine or digestive ointment, commonly made use of, for this will produce a considerable degree of inflammation in a short time.

CLYSTERS.

A Clyster is an injection,—a liquid substance injected into the lower intestines, for the purpose of promoting alvine discharges, relieving costiveness, and cleansing the bowels. A variety of compositions have been recommended for clysters, there being scarcely an article in the materia medica that has not been employed in this way. I have found, however, from considerable experience, that for a common clyster, water is as efficacious as the most elaborate composition. Where a purgative clyster is required, from four to eight ounces of common salt may be added; and if an anodyne be wanted, or an astringent, let half an ounce of opium be dissolved in a quart of water-gruel. If a clyster be employed for the purpose of emptying the large intestines, or of purging, the quantity of the liquid should not be less than a gallon or six quarts; but when used as an anodyne or astringent, from a quart to three pints will be sufficient.

THE PULSE.

This is indeed the key of medicine; for without authentic and minute information on the subject of the *pulse*, it is impossible for us to proceed to administer medicine to the sick animal with any certainty of success. When I talk about the pulse, I mean the beating or throbbing of an artery—there being no pulse whatever in the veins. The meaning of an *artery* is, a large blood-vessel, branching out into smaller ones, which carry the blood from the heart to the extremities of the body—in other words, to the points of the fingers and toes—where they join with the veins, which bring the blood back again to the heart. By pressing your finger hard on the artery, you will feel the pulse beat under it distinctly. Every time the heart beats it throws a volume of blood into the arteries; then again the heart contracts or draws up, and a fresh portion of blood is forced on into the arteries. Reflect for a moment on this wonderful machine, the heart: it goes with greater regularity than any watch, at the rate of about four thousand one hundred and fifty strokes every hour. This swelling and contracting of the artery, then, constitutes what I mean by the *pulse;* and, therefore, you may find the pulse in any part of the body where the artery runs near enough to the surface; for instance, at the fore leg, in the temple, under the lower jaw. Great advantage may be derived from attending to the state of the pulse in the management of sick horses, as we are thereby enabled to judge of the degree or violence of the disease, and the probability there may be of recovery; we are in some measure assisted also by it in ascertaining the nature of the complaint, and in the application of remedies. In a healthy horse the pulsation is about 36 or 40 per minute, and may be felt very distinctly on the left side, or in any artery that passes over the lower jaw-bone; in short, pulsation may be felt in every superficial artery. When the brain is oppressed the pulse generally becomes unusually slow. In a case of water on the brain which occurred lately, the pulse fell to 23 per minute. In cases where a horse appears dull, and refuses his food, it is advisable to examine the pulse, and if it be found to exceed the standard of health, his symptoms should be closely watched, for by timely interference many dangerous complaints may be prevented. When the pulse rises to 80 or 90 per minute the disease generally proves fatal.

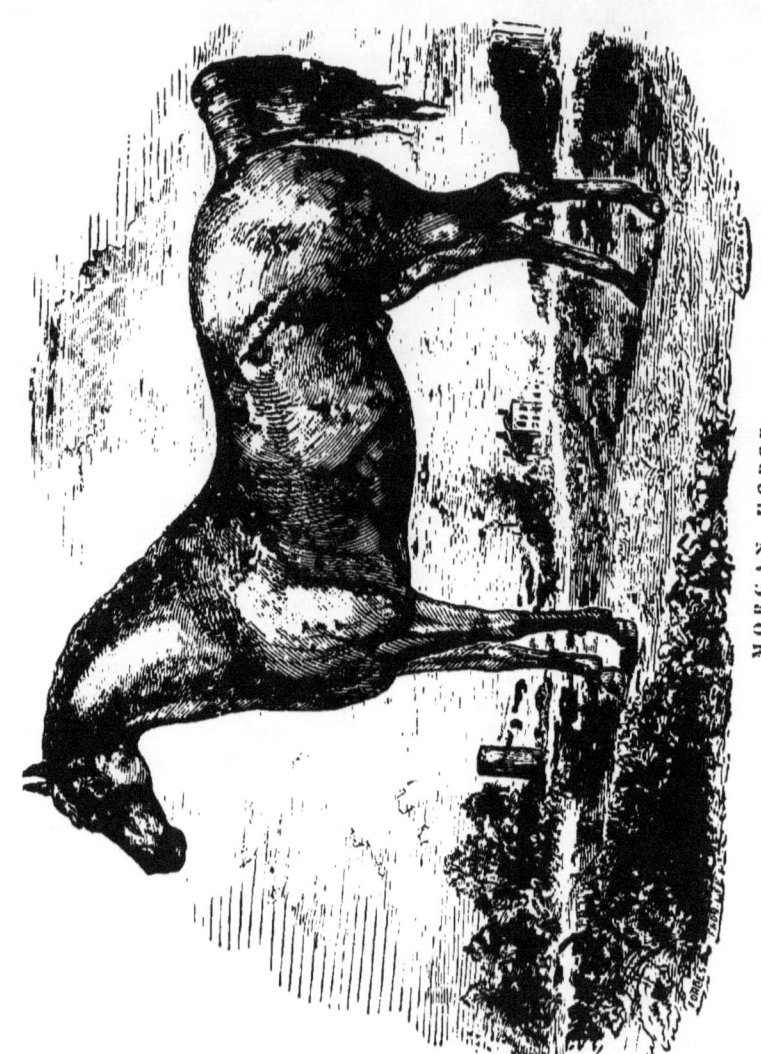

MORGAN HORSE.

GENERAL MANAGEMENT OF THE HORSE.

FEEDING, EXERCISE, AND GROOMING.

These subjects are of more importance than is commonly believed, and require more attention than is generally paid to them, as the health and condition of the horse depend greatly on his being properly managed. When in a state of nature, or when living under natural conditions, there is no doubt that green food, which then is bountifully provided for him, is better calculated than any other to keep him in perfect health, and satisfy his wants; but when he is domesticated, and employed in the various labors for which he is found so essentially useful, it is necessary to adapt the quantity of his food to the nature of the work he has to perform.

When a horse is to be got in condition, we should find out for what kind of labor he is designed—whether it be for the turf, the chase, or the road. A horse, without doubt, provided he is in a state of health, may have his condition and wind brought to the highest state of perfection it is capable of, merely by judicious management in respect to feeding, exercise, and grooming.

Notwithstanding the great mystery and secrecy affected by those who make a business of training race-horses, it is a very simple process, and easy to be accomplished by any one; and the whole mystery may be summed up in these three words—*feeding, exercise,* and *grooming*. It is a fact perhaps not sufficiently understood, that the strength of an animal, or any part of the body, may be increased to a considerable degree by means of exercise properly conducted; and as breathing is effected by muscular exertion, good wind depends on the strength of those muscles by which breathing is performed; and by keeping in view this single principle, we shall do more for the improvement of a horse's wind than we could by learning all the mysteries of training.

In order to have a clear idea of the method of getting the horse into high condition and good wind, let us suppose him just taken from grass. When a horse is taken from grass, or from his natural state, in order to be got in condition for racing, hunting, or the road, the first object should be to bring about the necessary change in his food and other circumstances as gradually, and with as little inconvenience to the animal as possible. When he is taken from grass, let him be put at first into a large, airy stable, and suffered to exercise himself in it. He should be allowed drink frequently, and, instead of depriving him suddenly of green food, allow him at first carrots, bran, and a moderate quantity of oats. He should be walked out at least once a day. His allowance of oats should be gradually increased, and that of bran and carrots in like manner diminished, until the latter are wholly discontinued. If he be a large drinker, he should be allowed but a moderate quantity at once; but at all times, and under all circumstances, it is proper to allow a horse water four times a day, which, instead of oppressing his stomach or injuring his wind, will facilitate digestion, and materially conduce to the preservation of health and improvement of condition. I am aware of the prejudice that exists against this practice—that it is supposed to give the horse a large belly, and render him unfit for galloping any length of time without endangering his wind. I am convinced, however, not only from my own experience, but by that of some experienced sportsmen also, that, so far from injuring a horse in any respect, it is extremely beneficial; and that when a horse is allowed to drink four or five times a day he is not inclined to drink much, and often does not drink so much in the twenty-four hours as one that is allowed to drink only twice a day as much as he pleases. As the horse's allowance of oats is increased, so should be his exercise; and, if this be properly managed, there will be no necessity for medicine or drugging. We should, however, watch the horse carefully during the time we are increasing his allowance of oats and diminishing that of green food, and if he appear dull, or have a cough, however trifling, it indicates an inflammatory disposition of the body, and points out the propriety of a laxative, more exercise, and less oats; but, under proper management, such symptoms will never take place, though they almost always do when a horse is changed from grass to a close stable and dry food too suddenly, and in such cases it may be advisable to to give a purgative medicine to prevent the occurrence of a serious disease. It is, perhaps, from this circumstance that the absurd custom of giving exactly three strong doses of physic, as a necessary preparative, took its origin. I have seen many valuable horses quickly destroyed by strong physic, and a great many have never perfectly recovered from the debility it occasioned. The first week that the horse is taken into the stable, walking exercise is the most proper, but after this it may be gradually increased to a trot or canter; and if the exercise occasions any degree of perspiration, he should be carefully cleaned, and otherwise attended to, as soon as he gets into the stable.

When we thus gradually bring a horse from a state of nature—that is, from the open air and green food—to a comfortable stable and dry grain, there will be no danger of those troublesome diseases which are often the consequence of sudden changes and a different kind of management; and by duly proportioning his exercise to the nutriment he receives, and by gradually bringing the muscular system to that degree of exertion for which the animal is wanted, there is no doubt but his wind, strength, activity, and general condition will be brought to the highest state of perfection it is capable of attaining. In describing the general management of horses in the stable, I think it necessary to be very particular, as there are many apparently trifling circumstances which have considerable influence on the horse's health, though generally little understood.

All horses that are obliged to undergo great and rapid exertion, such as drawing mail-stages, hunting, &c., at certain periods, require a different treatment from such as work more moderately. The former have occasion for lying down as much as possible, that the muscles may more rapidly recruit their strength; but the latter do not require so much rest in a recumbent state, and suffer no inconvenience from standing during the day; therefore the litter should be removed every morning and shook up in the open air. The advantage of this plan is considerable, though it may be thought by knowing grooms to be an unnecessary trouble. By this means the feet will be kept cool, and the hoofs will not be so disposed to contract and shrink; for straw, being a bad conductor of heat, causes the feet to become too hot, in which state the horny matter has always a tendency to contract; hence arises the thrush, sand-cracks, &c. Horses that are constantly on the floor should have their feet stopped with a mixture of cowdung and blue clay. The feet should be examined daily, and if the soles appear to be softened too much—that is, if the horn bends or gives way under the pressure of the thumb—the stopping must be discontinued.

Horses that work hard should have plenty of oats and hay, with a moderate quantity of corn or beans. The latter, however, should not be allowed unless the horse's work be considerable, as under moderate exertion they dispose the system to inflammatory complaints, such as coughs, inflamed eyes, &c. Horses whose labor is severe are often injured by being stinted in water, particularly when they are allowed a large quantity of food. It is a common practice with wagoners, when their horses come in from a long and fatiguing journey, their strength almost exhausted by long and continued exertion, to offer them immediately an unlimited quantity of food, and very little (and sometimes not a drop) of water; consequently the stomach is not able to digest the food that is taken in, and the staggers are often the consequence of such management. When a horse comes in from his work, and after he is sufficiently cool, he should always be allowed a small quantity of water before he is fed; and if he be allowed a little immediately after feeding, instead of injuring the animal it will promote digestion, and prove beneficial. It is certainly a good plan to give horses a moderate quantity of water before the end of the journey; and I am satisfied that, by allowing them to sip a little water several times during a long journey, particularly in warm weather, they are refreshed and invigorated, but never injured. When beans are given to a horse, they always should be broken and mixed with oats. A horse of medium size that is moderately worked does not require more than a peck of good oats, and about twelve or fourteen pounds of hay, in twenty-four hours; but large draft-horses require a greater quantity of both hay and oats.

All horses that are employed for expeditious traveling require great attention, as regards feeding, grooming, &c. Their allowance of hay should not exceed twelve pounds in twenty-four hours, and that should be divided into three feeds—four pounds in the morning, two at noon, and the remainder at night. If a peck of oats be allowed for the same period, it should be divided into at least three feeds, giving water before each.

It has been asserted by some that horses work better, and more effectually preserve their wind and condition, when allowed only a small quantity of water, or, as they express it, "it matters not how little they drink, provided they feed heartily." This opinion, like many others, has arisen from the foolish and absurd practice of forming general rules upon a few facts or a very limited experience, and too often from examining those facts through the medium of prejudice. I will admit that we sometimes meet with horses that become loose in their bowels, and fall off in condition, sweating violently, and appearing much fatigued from moderate exercise, if allowed to drink even two pails (five or six gallons) in twenty-four hours, particularly when employed in any kind of violent exercise; but this is to be attributed to a weakness of constitution not often met with in horses, and points out the necessity of observing a horse attentively when we first undertake the management of him. If we find a horse shivering, and his coat staring immediately after drinking freely, it is not to be hastily concluded that he is to be allowed only a small quantity of water daily. In such a case a moderate quantity should be given at once, and the horse should be exercised immediately after, in which way he will generally be brought to drink a proper quantity in the course of the day without inconvenience.

Horses that are idle should be exercised regularly every day; the best time for exercising the horse is early in the morning, as soon as the stable is opened, during which time the stable door should be kept open, and all the foul litter thrown out. A small quantity of straw should be spread under the horse to prevent the horse from splashing his legs in staling. Horses of a full habit, or such as are subject to humors, are greatly benefited by exercise, which, on such occasions, may be carried so far as to produce sweating. But great care is then necessary; they should be walked about for some time, that they may cool gradually; and as soon as they return to the stable they should be wiped dry, and their legs well rubbed. Swelling of the legs, grease, inflamed eyes, and other troublesome complaints, will be thus more effectually prevented than by bleeding or physic, which, though it may give temporary relief, will gradually increase the disposition to disease.

As a substitute for exercise, the horse may be kept in a loose stall; if the stall be sufficiently large, he can turn himself about and enjoy comparatively a state of liberty.

Whenever the weather is temperate, horses may be cleaned in the open air. The common practice of washing the legs with cold water should never be allowed, unless the horse be exercised, or have his legs well rubbed immediately after. It is not only absurd but dangerous to plunge a horse into a river while sweating from severe exercise,—a practice commonly adopted by proprietors of stage horses; that it is often done with impunity must be granted, but it is probable that many of them suffer from the treatment, though the ill effects are not often immediately observed.

When a horse returns from exercise or work, his feet should be carefully pricked out and washed, and if the hoof be dry and brittle, feeling hot and appearing contracted, a mixture of cowdung and soft clay should be applied to the sole.

We should remember that when a horse is changing his coat, that is about the latter end of September and beginning of October, he is more susceptible of cold than at any other time; and as then the coat falls off readily, the curry-comb should be laid aside, and the horse exposed as little as possible to cold or rain. Moderately warm clothing, and frequent hand-rubbing to the legs, will be found highly useful at this time. When these precautions are neglected, horses often become weak and unfit for much work, sweating profusely from moderate exercise, and sometimes purging; troublesome cough and staring coat generally accompany these symptoms. The common remedies on this occasion are bleeding, or strong purgatives, which are sure to increase the debility; nor are antimonials or medicines that act upon the skin proper to be given. The most effectual medicines to be given in these cases are tonics, with moderate stimulants. For particulars, see Tonics and Stimulants, which may be found under their proper heads.

THE AGE OF A HORSE.

The age of a horse may be told by certain marks in the front teeth of the lower jaw, and the tusks, until about the eighth year, at which time they are generally worn out. An experienced person can, however, after this period, judge of

GENERAL MANAGEMENT OF THE HORSE. 161

the age with some degree of accuracy, by the countenance and general appearance of the animal, as well as by the length and form of the tusks.

A colt begins to change his *sucking teeth* between the second and third year. The *sucking* teeth, as they are termed, are small, and of a delicate color, and some of them are perfectly smooth on the upper surface; others have a small narrow cavity on that surface, but very unlike those marks of the permanent teeth by which we judge of the age. The horse has twelve teeth in the front of the mouth, six in the upper and six in the lower jaw. (We take no notice of the molars, or grinders, as they are not connected with the subject.) When a colt is three years old we may observe that the four front *sucking teeth* are lost, and that, instead of them, four others having sprung up, of a very different appearance, being larger, of a darker color, and having a considerable cavity on the upper surface, and a small, dark-colored groove in front; these are termed *horse or permanent teeth*. Between the third and fourth year the four teeth next to these are lost, and replaced, in the way we have just described, by horse teeth; so that when a colt has completed his fourth year, there are eight horse teeth observable, and only four colt's teeth—one at each extremity, or corner, as it is termed. About the middle of the fifth year these also fall out, and are succeeded by horse teeth. The corner teeth of the horse, and particularly of the under jaw, are different from the rest, being smaller and of a shell-like appearance; these cavities are chiefly within—the upper surface being a mere edge; but about the end of the fifth year they are larger and more like the other teeth.

The tusks make their appearance between the fourth and fifth year, generally, though sometimes earlier. They are four in number, and situated about an inch from the corner teeth; at first they are small, terminate in a sharp point, are rather convexed on their external surface, but within have two concavities, or grooves, separated by a ridge. These, as well as the teeth, are gradually undergoing an alteration in their form, becoming longer, and losing the concavities on the internal surface. About the seventh year the concavity is considerably diminished, and in old horses the surface becomes convex, and the tusk acquires a round form, and the extremity, instead of being sharp, is quite blunt, as if the point had been broken off, and the new surface afterward polished. We must now return to the teeth, the appearance of which we have described as far as the completion of the fifth year. After this period we judge of the age by the size of those cavities which we have described, on the upper surface of the teeth; for the friction to which that surface is almost constantly exposed gradually wears it down, and at length the cavity or mark is totally obliterated. The marks in the upper teeth most commonly remain until the twelfth year, sometimes longer, but those in the under teeth are worn out about the end of the eighth year; we shall therefore confine our description now to the under jaw.

The two front teeth being the first that make their appearance, it is obvious that their marks sooner wear out than those of the other teeth; and if we examine the mouth of the horse that has just completed his fifth year, we shall find that they are nearly, and sometimes quite, worn out; those in the adjoining teeth are about half their original size, while the marks of the corner or end teeth are perfect. At the end of the sixth year, the only cavities observable are in the corner teeth, and these are about half their original size, the tooth has at this period lost the shell-like appearance we have before described, and is not different from the other teeth, except in having a mark or cavity on its upper surface. At the end of the seventh year, the marks of the corner teeth are also obliterated, and then the horse is said to be aged. It is often the case, however, that the marks of the corner teeth are not totally effaced at this period; a small dark-colored spot may be observed in most horses, until about the end of the eighth year. From this period we have no criterion by which the age of a horse may be ascertained, but there are marks on the upper teeth that will enable us to judge of the age until the thirteenth year; the marks of the front teeth being worn out when he becomes eight years old, those of the adjoining teeth at ten, and the corner teeth at twelve; but these marks cannot always be depended upon.

CRIB-BITING.

Crib-biting is a very serious habit. The horse lays hold of the manger with his teeth, violently extends his neck, and then, after some convulsive action of the throat, a slight grunting is heard, accompanied with a sucking or drawing in of air. It is not an effort at simple eructation, arising from indigestion. It is the inhalation of air, produced by a convulsive action of the throat.

Effects.—The teeth are injured and worn away, giving the horse an old appearance. A considerable quantity of corn is lost, for the horse will frequently crib when eating, with his mouth full, and a greater part of the corn will fall over the edge of the manger. Much saliva is lost in the act, which must be of serious consequence in impairing digestion. The crib-biting horse is more subject to colic than other horses, and to a species difficult of treatment and frequently dangerous. Although many a crib-biter is stout and strong, and capable of all ordinary work, those horses do not generally carry so much flesh as others, and have not their endurance. On this account crib-biting horses have been decided to be unsound; so a case was decided before Lord Tenterden.

Causes.—The causes of crib-biting are various, and some of them beyond the control of the proprietor of horses. It is often the result of imitation; but it is more frequently the consequence of idleness. The high-fed and spirited horse must be in mischief; if he is not usefully employed. It may be caused by partial starvation, whether in a bad straw-yard, or from unpalatable food. This trick is exceedingly contagious. Every horse in the stable is likely to acquire the habit.

Remedy.—Some have recommended turning the horse out for six months, but this has never succeeded except with a young horse, and then rarely. The old crib-biter will employ the gate for the same purpose as the edge of the manger, and we have often seen him galloping across a field for the mere object of having a gripe at a rail. Some have recommended muzzling the animal sufficiently tight to prevent him grasping the edge of the manger, but loose enough to let him pick up his corn and pull his hay. Others have recommended a strap buckled tight around the neck. There have been numerous ways invented to break the horse of this habit, but no one has been successful, and all have been more or less inconvenient to the horse, until my invention. This is safe, sure and effectual, and is no inconvenience to the horse whatever. It consists simply in this: The stall should be sealed up in front and each side, as high as the chamber floor, and as far back as the horse can reach; then make a common box-manger, and over the top of the edge of the manger place a roller six inches in diameter, long enough to reach the whole length of the manger, fastened at each end by a pivot, so that it will roll perfectly easy. Here you have the plan: The roller is so large that the horse cannot grasp it, and it being placed over the edge of the top of the manger, he cannot get hold of that; and if he attempts to crib by bearing down on the roller with his upper teeth, it will roll out from under the pressure, as the horse always draws towards him as he presses down; this is a very common way of cribbing. I once purchased a valuable stallion that had this habit to such a degree that he could not be got in condition, and would fill himself so full of wind in fifteen or twenty minutes after being put into the stable, as to throw himself into the colic. I invented this plan for his benefit, which was successful in breaking him of the habit. He was then seven years old, but had never developed as a full-grown horse. After breaking him of the habit he commenced improving, and in one year's time weighed one hundred pounds more than when I purchased him, and never again acquired the habit whilst I knew him.

WIND-SUCKING.

This habit is similar to that of crib-biting. It arises from the same causes, the same purpose is accomplished, and the same results follow. The horse stands with his neck bent; his head drawn inward; his lips alternately a little opened and then closed; and a noise is heard as if he were sucking. There is in this case the same want of condition and the flatulence which I have described under the last head. This diminishes the value of the horse almost as much as crib-biting; it is as contagious, and as inveterate. The only remedies, and these will seldom avail, are tying the head up except when the horse is feeding, or putting on a muzzle with sharp spikes towards the neck, and which will prick him whenever he attempts to bend in his head for the purpose of wind-sucking.

THE SECRET OF HORSE-TAMING.

As Rarey's secret of horse-taming by mechanical processes has been exposed and made known through the public journals, I shall not dwell on his method. It is well-known to all that have practiced it, that his art is sadly deficient; and that to control the Horse, he must be affected through the medium of his senses, and the sense of smell is the most effectual. "The Horse is unlike the dog, the bull, and most other quadrupeds, in two respects, both of which peculiarities run into one tendency. The Horse has no weapons of defence, and hence is more dependent than other animals on his sense of smell for his protection. Indeed, it is scarcely known how very keen this sense is in the Horse, and how much he depends upon it always. In one of the earliest allusions to the Horse (book of Job, xxxix. 20), this trait in his nature is noticed, when the writer speaks of the 'terrible glory of his nostrils,' and declares that he *smells* the battle afar off."

It is remarked that the Horse, unlike other animals, breathes only through his nostrils, and not through the mouth, like the ox and the dog. And this fact goes to confirm the view here given as to the strength and importance of this function of smell, as if the breath of the Horse had been confined to the nose to keep constantly active the sense of smell, which the horse needs, not merely for the selection of his food, but also for calling into exercise his caution and combativeness,—those functions upon which his protection and life so very much depend. The Horse is, therefore, excited, alarmed, and repelled through his sense of smell more than through his sense of hearing, feeling, tasting, or sight. He is repelled through each of his senses at times, but always and most through his sense of smell, no matter what the object may be by which he is frightened.

A knowledge of the peculiar function of smell in the Horse suggests at once what the philosophy of horse-taming must of course include; for if the Horse is so much repelled and frightened through the sense of smell, it is easy to see that he may be equally attracted through the sense of smell, if we can only approach him with those substances which will sufficienty gratify it. The scent by which he is attracted should be sufficiently strong and agreeable to him to overcome the smell of all other objects, animate or inanimate, which excite his fear, and are disagreeable to him.

It is said that each human being "has his price," and may be bought or induced to do certain things—that is, has certain tastes, certain senses; and when these senses are gratified more or less, we are more or less pleased. And so of all animals; they may be influenced by addressing their strongest senses. And when, as in this case of the Horse, it is the sense of smell, and that sense is very strong, we must find those scents which gratify him most, and such as will thus absorb and annihilate all other scents that are disagreeable. For this purpose, you will find the oil of *rhodium* and oil of *cummin* to be the most effectual. Take equal parts, mix and keep in air-tight phials ready for use. These should be rubbed on the inside of the nostril. In a short time you will have the horse under your control, and he will follow you by the sense of smell. The wildest horse may be caught by cornering him where he cannot escape. Rub your hands with the oils cummin and rhodium; have your pole, with a small piece of cloth wound on the end, which has been saturated with the oils. Approach him from the windward, and you may thus attract him, even before he is within the reach of your pole. Proceed gently, until you can reach his back with the end of your pole, precisely in the same way as if your arm were elongated to the length of your pole; and you pat him and work and move the pole over his back, gradually and gently approaching him, so that he is attracted by the sense of smell of the oils on your hands, and you can shortly rub some of the oils on his nose; this once done, you can secure him at your pleasure. After a horse is once caught in this way, you will find no trouble in catching him. A failure of a few times should not discourage you—repeat the process until you succeed. With some horses you may succeed best by dropping some of the oils upon an apple or a lump of sugar, and give him to eat, and breathe into his nostrils; this will often work like a "charm." But then it should be borne in mind that there is a difference in horses as really as in human beings. Horses that have large caution or fear, it is, of course, much more difficult to control. But the agreeable excitement of the sense of smell overcomes the sense of fear, and fear once subdued, it enables him to do your bidding.

Make your horse love you; attract him by all means in your power. How you should reach him through the sense of smell (always when you fail in other methods) I have told you: the nearer you get to him the better. I once heard the celebrated Catlin say that he had seen the little calves of the buffalo, on the western prairies, following the hunters after their dams had been killed. To induce the calves to follow them, the hunters breathed into their nostrils and fondled them. Thus, if your person is scented with the oils of cummin and rhodium, or you breathe into the nostrils of your horse, you pathetize him with your sphere. Fondle, or, as the Scotch would say, cuddle him; thus he becomes acquainted with you; he associates you with what he loves to smell or eat; and thus you gain your power over him. Do not be cross or cruel; do not beat him; but caress, love him, fondle over him, and thus cause your horse to love you. Love is the charm—the great secret—and without it you can never control a vicious horse. It must be in you; must come out of you; must appear in your words, tones, and all your actions, if you wish the horse to love you. There are many good-natured, kind people, who would be glad to make their animals love them, but they lack wisdom; they do not seem to appreciate those means which alone are appropriate for securing this result.

These oils, cummin and rhodium, are the celebrated Arabian horse-taming secret or receipt, that has often been sold for fifty dollars, with the knowledge of its use.

PRESCRIPTIONS.

POWDERS.

Stimulating Powder, No. 1.

Powdered capsicum, ... ½ ounce.
Common salt, . . . "
Mix. Dose, a table-spoonful.

Cathartic Powder, No. 2.

Powdered aloes, ... 1 ounce.
" blood-root, . . . 1 drachm.
" lobelia, ... 3 "
Mix. To be given at one dose.

Expectorant Powder, No. 3.

Powdered liquorice, ... 4 ounces.
" blood-root, . . . ¼ "
" slippery elm, . . . 8 "
' pleurisy-root, . . 8 "
Mix. Dose, one tablespoonful.

Expectorant Powder, No. 4.

Powdered ipecac, . . . 2 ounces.
" lobelia, . . . 2 "
" blood-root, . . . 3 "
" liquorice, . . . 4 "
Mix. Dose, from one to two ounces.

Tonic Powder, No. 5.

Powdered ginger, ... 2 ounces.
" golden seal, . . . 1 "
" poplar bark, . . . 2 "
" capsicum, . . . 1 drachm.
Common salt, ... ½ pound.
Mix. Dose, one ounce, to be given in all cases where a stimulant and tonic is required.

ANODYNE POWDER, No. 6.

Powdered gum benzoin,	2 ounces.
" opium,	$\frac{1}{4}$ "
" liquorice,	8 "
" anise seed,	2 "
Ginger,	4 "

Mix. Dose, one ounce. This is one of the best anodynes in use, and is applicable in all cases of cough that are produced by irritation.

COMPOSITION POWDER, No. 7.

Powdered bayberry bark,	6 pounds.
" ginger,	3 "
" cloves,	6 ounces.
" capsicum,	4 "
" cinnamon,	3 "

Mix. Dose, one ounce. This is one of the best stimulating powders for colds, or in any case where sweating is indicated, that I ever made use of. Dose, one ounce, to be steeped in a pint of hot water and given as a drench.

FEVER POWDER, No. 8.

Powdered white root,	6 ounces.
" skunk cabbage,	4 "
" lobelia herb,	4 "

Mix. Dose, one ounce.

DIAPHORETIC POWDER, No. 9.

Powdered gum camphor,	2 drachms.
" opium,	$\frac{1}{2}$ "
" ipecac,	1 "
Cream of tartar,	1 ounce.

Mix. Dose half ounce.

CATHARTIC POWDER, No. 10.

Powdered aloes,	4 ounces.
" rhubarb,	2 "
" gamboge,	1 "
" capsicum,	$\frac{1}{2}$ "
" lobelia,	$\frac{1}{2}$ "

Mix. Dose from one to two ounces.

Tonic Powder, No. 11.

Powdered unicorn root, . . . 8 ounces.
" golden seal, . . 4 "

Mix. Dose one ounce.

DRENCHES.

Diarrhea Drench, No. 1.

Pulverized gum opium, . . . 4 ounces
" " kino, . . 1 "
" " camphor, . . ½ "
" nutmeg, . . . 6 "
Alcohol, ... 1 quart.

After it has stood nine days, it is fit for use. Dose, an ounce as a drench in one pint of water.

Cathartic Drench, No. 2.

Pulverized rhubarb, . . . 1 lb.
" gum arabic, . . 1 "
" aloes, ... 1 "
Oil anise-seed, ... 1 ounce.
Whisky, ... 1 gallon.
Molasses, ... 2 quarts.

Tincture the articles in whisky for nine days, then add the molasses, and it is fit for use. Dose, four ounces.

Stimulating Drench, No. 3.

Pulverized capsicum, . . . ½ ounce.
Common salt, ... 3 "

To be given as a drench, in a pint of gruel. This should be given where there is an inactive state of the stomach or bowels.

Diuretic Drench, No. 3.

Balsam copaiba, ... ½ ounce.
Sweet spirits of nitre, . . 2 drachms.
Flaxseed tea, ... 1 pint.

To be given at once.

PRESCRIPTIONS.

Colic Drench, No. 4.

Tincture of ginger, ... 1 ounce.
Paregoric, ... 4 "
To be given at a dose.

Stimulating Cathartic Drench, No. 5.

Linseed oil, ... 12 ounces.
Tincture of aloes, . . . 2 "
" ginger, ... 1 "
Give the whole at a time in a pint of water.

Nauseating Drench, No. 6.

Pulverized aloes, ... 6 drachms.
" lobelia, . . . 2 "
Warm water, ... 1 pint.
To be given at a drench.

Colic Drench, No. 7.

Pulverized capsicum, . . . $\frac{1}{2}$ ounce.
Spirits of camphor, . . . $\frac{1}{2}$ "
Whisky, ... 1 pint.
To be given in a pint of water.

OINTMENTS.

Magnetic Ointment, No. 1.

Tobacco, ... 1 lb.
Raisins, ... $\frac{1}{2}$ "
Hard cider, ... 1 gallon.

Evaporate to one half in a covered vessel, strain and add four ounces tincture of lobelia, and six pounds of lard, evaporating slowly to an ointment. This is the celebrated magnetic ointment, and is one of the most valuable ointments in use.

Egg Ointment, No. 2.

The yolks of . . . 3 eggs.
Common salt, . . . 3 tablespoonfuls.

Mix well, and rub well on the inflamed parts. This is a valuable ointment for inflammation.

Ointment for Stiff Joints, or Relaxing Ointment, No. 3.

White-pine bark	4 ounces.
Bitter-sweet bark, from the roots,	4 "
Wormwood herb	2 "
Chamomile,	1 "
Hearts of mullen,	2 "
Lard,	2 lbs.

Put the whole in an earthen vessel, and let them simmer over a slow fire until the strength is extracted; then strain it, and it is fit for use. This is truly a valuable ointment, and this receipt has been sold for twenty-five dollars, to my knowledge.

Spavin Ointment, No. 4.

Mercurial ointment,	1 ounce.
Iodine, "	1 "
Gum camphor,	2 "
Oil of origanum,	6 "

Mix and apply once a day.

Spavin Ointment, No. 5.

Spanish flies, powdered,	½ ounce.
Oil of turpentine,	1 "
Hogs' lard,	4 "

Spavin Ointment, No. 6.

Common tar,	4 ounces.
Vitriolic acid,	1 drachm.
Origanum,	½ ounce.
Hog's lard,	2 "
Spanish flies, pulverized,	½ "

Add the vitriolic acid gradually to the tar, and then the rest of the ingredients. It is necessary to be careful in mixing the vitriolic acid to the tar, for unless it is well mixed with the tar the acid will act as a caustic upon the skin and produce ulceration. This ointment is remarkably useful in removing enlargements of any kind, as well as spavin.

Brown Ointment, No. 5.

Fresh butter,	3 ounces.
White wax,	½ "
Red precipitate,	2 drachms.
Ointment of oxyd of tin,	1 "
Camphor, dissolved in sweet oil,	1 "

Mix the first two articles, and when nearly cold stir in the precipitate finely pulverized. This is an excellent ointment for inflammation of the eyes, grease, and scratches, or any old sore.

LINIMENTS.
Liniment, No. 1.

Alcohol,	1 pint.
Origanum oil,	2 ounces.
Gum camphor,	1 "
Aqua ammonia,	2 "
Sweet oil,	2 "

This is a good liniment, and is general in its application.

Laxative Liniment, No. 2.

Tobacco,	1 ounce.
Lobelia seed,	1 "
Capsicum,	½ "
Whisky,	1 quart.

Let it stand and tincture nine days, filter and add

Aqua ammonia,	1 ounce,
Oil origanum,	1 "
" cedar,	1 "
Sweet oil,	1 "

and it is fit for use. This is a good liniment to relax stiff cords and joints.

Stimulating Liniment, No. 3.

Oil cedar,	1 ounce.
Tincture of capsicum,	6 "
Linseed oil,	1 pint.

This is one of the best liniments in use for rheumatism or sprains.

Stimulating Liniment, No. 4.

Tincture of balm of gilead,	8 ounces.
Oil of cedar,	1 "
Alcohol,	1 pint.

Horse Liniment, No. 5.

Spirits turpentine,	1 pint.
" camphor,	2 "
Origanum,	2 "
Oil succini,	2 "
" tar,	2 "
Barbadoes tar,	4 "
Laudanum,	2 "

Soap Liniment, No. 6.

Castile soap,	4 ounces
Gum camphor,	2 "
Oil rosemary,	$\frac{1}{2}$ "
Alcohol,	3 pints.

Volatile Liniment, No. 7.

Sweet oil,	2 ounces.
Aqua ammonia,	$\frac{1}{2}$ "

Shake well, and keep corked tight, and it is fit for use.

PILLS.

Pill Mass No. 1.

Pulverized aloes,	4 ounces.
" colocynth,	2 "
" gamboge,	2 "
Extract gentian,	4 "
Castile soap,	2 "
Oil cloves,	20 drops.

Mix, and divide into balls the size of a butternut. Dose, from one to three.

Pill Mass No. 2.

Pulverized aloes,	. . .	8 ounces.
" sal nitre,	. .	2 "
" gum myrrh,	. .	1 "
" cinnamon,	. .	1 "
" cloves,	. .	1 "
" ginger,	. .	1 "

Syrup sufficient to form a mass suitable to pill. To be divided into twenty pills. Dose, from four to five.

Pill Mass No. 3.

Pulverized gum gamboge,	. .	5 ounces
" aloes,	. .	8 "
" scammony,	. .	2 "
Sal nitre,	. . .	1 "
Gum arabic,	. . .	4 "

Warm water sufficient to form a consistency for pilling. be divided into forty pills. Dose, from three to five.

BALSAMS.

Cough Balsam, No. 1.

Molasses,	. . .	1 pint.
Tincture lobelia,	. .	4 ounces.
Essence peppermint,	. .	1 "
" anise seed,	. .	2 "

Mix. Dose, half ounce.

Cough Balsam, No. 2.

Molasses,	. . . .	1 quart.
Castor oil,	. . .	4 ounces.
Paregoric,	. . . .	8 "
Spirits camphor,	. .	4 "
Essence peppermint,	. .	2 "

Dose, one ounce.

Restorative Balsam, No. 3.

Sweet oil,	. . . .	2 quarts.
Spirits turpentine,	. .	½ pint.
Oil origanum,	. . .	4 ounces.

Mix. Apply once or twice a day, and bathe it in well with a

hot brick. This one of the best mixtures for rheumatism that I ever used.

VERMIFUGES.

Vermifuge No. 1.

Powdered aloes,	2 ounces.
" wormseed,	2 "
" ginger,	2 "
" poplar bark,	2 "
Common salt,	2 "

Mix. Dose, one ounce night and morning.

Vermifuge No. 2.

Castor oil,	12 ounces.
Wormseed oil,	$\frac{1}{2}$ "
Tansy oil,	3 drachms.

To be given at once in linseed tea.

Vermifuge No. 3.

Wormseed oil,	1 drachm.
White sugar,	3 "
Gum arabic,	2 "
Mint water,	1 pint.

To be given as a drench.

DIURETICS.

Diuretic No. 1.

Balsam copaiba,	1 ounce.
Spirits nitre,	2 "
Oil almonds,	2 "
Spirits turpentine,	1 "
Gum camphor,	1 scruple.

Mix. Dose, one ounce.

Diuretic No. 2.

Fluid extract of dandelion,	2 ounces.
" " of sarsaparilla,	2 "
" " of Indian hemp (*apocynum cannabinum*),	1 "

Mix (Dose) two drachms in a pint of flax-seed tea. This is

one of the best diuretics that I ever used in all dropsical affections or swelling of the legs, and is safe and effectual in all cases where a diuretic is required; it operates on the secretions in general, and is a good alterative as well as diuretic.

TINCTURES.

SUDORIFIC TINCTURE.

Powdered ipecac,	1 ounce.
Saffron,	1 "
Camphor gum,	1 "
Virginia snake root,	1 "
Opium,	1 "
Diluted alcohol,	3 pints.

This is one of the best sudorific or sweating medicines known, and comes as near being a specific as any medicine in use. Dose—from half to one ounce.

RHEUMATIC TINCTURE, No. 2.

Alcohol,	1 pint.
Oil cloves,	1 ounce.
" peppermint,	½ "
" hemlock,	1 "
" origanum,	1 "

PAIN-KILLING TINCTURE, No. 3.

Alcohol,	1 pint.
Oil cloves,	1 ounce.
" cinnamon,	1 "
" anise,	1 "
' origanum,	1 "

STIMULATING TINCTURE, No. 4.

Oil cedar,	4 ounces.
" sassafras,	2 "
Capsicum,	2 "
Alcohol,	1 quart.

Cholera Tincture, No. 6.

Loaf sugar,	2 ounces.
Cloves,	$\frac{1}{2}$ "
Cayenne pepper,	$\frac{1}{4}$ "
Gum camphor,	$\frac{1}{2}$ "
Nutmeg,	1 "
Brandy,	1 pint.

Nerve Tincture, No. 7.

Powdered nerve-root,	4 ounces.
" fennel seed,	8 "
" anise "	6 "
" Virginia snake root,	4 "
Whisky,	1 gallon.

Nerve Tincture, No. 5.

Powdered nerve-root,	4 ounces.
" valerian,	4 "
" scullcap,	4 "
Diluted alcohol,	1 quart.

Dose, one half-pint.

INJECTIONS OR ENEMAS.

Injection No. 1.

Powdered aloes,	4 drachms.
Common salt,	1 "
Hot water,	1 quart.

To be given blood-warm.

Injection No. 2.

Soft soap,	2 ounces.
Common salt,	1 "
Hot water,	1 quart.

Injection No. 3.

Castor oil,	4 ounces.
Flax-seed tea,	1 pint.

STIMULATING INJECTION, No. 4.

Capsicum, ... 1 drachm.
Common salt, ... 1 ounce.
Hot water, ... 1 quart.

ANODYNE INJECTION, No. 5.

Pulverized slippery elm, . . 1 ounce.
Morphine, ... 1 scruple.
Warm water, ... 1 quart.

This should given in all cases of diarrhea or irritation of the bowels.

DISEASES OF HORNED CATTLE.

I HAVE found by experience and observation that the diseases of horned cattle are few and simple. When animals live their wild and natural state, they are seldom or never sick; but when deprived of all natural and healthy conditions, by shutting them up and depriving them of pure air, water, exercise, and their proper food, they get sick, the same as we do; and we have veterinary surgeons, cow-doctors, &c., to cure them. As cattle live under more natural conditions, and are not compelled to transgress the law of their nature so frequently as the Horse, they are not so subject to disease. All the cattle-doctors have treated on a variety of diseases,—diseases, that exist nowhere except in the imaginations of the authors. They, too, like the horse-doctor, have drawn their conclusions from medical works on the human subject. They have supposed that animals were subject to nearly all the diseases incident to mankind. Hence their multiplicity of disease. But their books must be filled up; they must be scientifically written, having the appearance of wisdom, at least. But a practical work is what the people want; a work based upon observation and experience—a practice that will apply in the diseases of their animals. My experience teaches me that there are but few diseases among cattle; which diseases will be treated of in a plain and comprehensive manner. In doing so, I shall lay down general principles upon which all diseases must be treated. I care nothing about names of diseases, if I only understand their nature and cause. By understanding the laws that govern in life and health, disease and death, we can treat disease successfully without going into the minutiæ of diagnosis or discrimination of disease.

at.

MILK FEVER.

This disease makes its appearance in one or two days after calving.

Cause.—This disorder happens when the cow is in good condition and full of blood. It may occur at any time of the year, but is not so common when kept on dry food, as in hot weather when the cow runs at grass.

Preventive.—It might be prevented by milking the cow before calving, and keeping on a low diet. It never happens to a cow with her first calf, and seldom with the second, and chiefly to great milkers, for few others have it.

After a cow has had this fever once, she is very liable to have it again; and there is little hope of her recovering from the second attack; consequently it would not be good economy to expose a cow the second time to the disease.

Symptoms.—The cow starts, staggers, and trembles till she comes down. The legs will soon become stiff and cold. She will soon have a quick, strong pulse, and a high fever; therefore prevent her being kept too hot in summer or too cold in winter.

Treatment.—This will depend chiefly on the symptoms and the stage or development of the disease. If the extremities are cold, apply the foot-bath—the manner of doing it may be seen under its proper head—and give a composition powder made as follows:

Ginger,	1 ounce
Capsicum,	$\frac{1}{4}$ "
Lobelia herb,	$\frac{1}{2}$ "
Boiling water,	2 quarts.

Give half of this when cool; and, in an hour, if this does not equalize the circulation, and the extremities are not warm, give the remainder; after this object is accomplished, give physic composed of

Mandrake, powdered,	1 ounce.
Cream of tartar,	2 "

If this does not operate in six hours, repeat the dose. The operation of physic should be assisted by the following injection:

Soft soap,	2 ounces.
Common salt,	1 "
Hot water,	1 quart.

To be given milk-warm, and repeated in short intervals until the desired effect is produced. After the bowels become sufficiently relaxed, diuretics should be freely given. This will be found to be a very convenient and effectual diuretic drink:—

  Clivers, ... 2 ounces.
  Juniper-berries, . . . 1 pound.

Bruise them, and put them into a jar, and pour on four quarts of boiling water; let them stand until cool, then strain, and give a pint every two hours, in four quarts of cold water, until a considerable diuretic effect is produced. For a change of drink, give linseed-tea or barley-water; if she refuses these, give plenty of water, moderately warm, in small quantities.

It is sometimes necessary to give fever-medicine. The following is very effectual:—

  Powdered whiteroot, . . 2 ounces.
   " skunk cabbage, . 1 "
   " lobelia herb, . . 2 "
  Boiling water, . . . 4 quarts.

If the fever runs high, give a pint of the above decoction, or tea, once in six hours until the fever abates. The cow should be placed upon plenty of straw, and where there is plenty of room, and mostly on the milk side; and draw the paps often to get what milk you can.

Never offer to get a cow up till she is able to stand. Some are brought so low by the disease as not to be able to stand on their feet in less than three weeks.

The cow should be turned over every two or three hours after the first day.

When the fever has abated, she will want nourishment; give good gruel frequently, and in small quantities. It is sometimes necessary to give a tonic when the cow gets very much reduced; for this purpose give

  Powdered ginger, . . . 2 ounces.
   " golden seal, . . 1 "
  Common salt, . . . 2 "

Dose, one ounce in wheat bran, or bran and water, three times a day.

Frequent hand-rubbing to the legs is of considerable importance, and, if they incline to be cold, rub them with the following liniment:—

Oil of cedar, . . . 1 ounce.
Tincture of capsicum, . . 6 "
Alcohol, ... 1 pint.

The back should also be rubbed with the same once a day. Manage as directed, and the cure is sure in nineteen cases out of twenty.

There are more cows that die of this than all other diseases put together, and especially among the high breed or blooded cattle; they are generally high fed, and stabled a good part of the year.

SIMPLE FEVER.

I have been able to discover but three kinds of fever in cattle, or three different developments of fever; first—the milk, or puerperal fever; second—simple fever; third—symptomatic fever.

Cause.—Simple fever is caused by taking cold, causing a collapse of the perspiring vessels. This will obstruct the circulation, causing that derangement in the system which brings on the simple fever. The simple fever is not so common as the symptomatic, nor so fatal; yet it is necessary to attend to it in season, or the unequal circulation of the blood will cause congestion and inflammation in some of the internal organs, causing a dangerous disease.

Symptoms.—Shivering; quick and full pulse; cold extremities; after reaction takes place, the heat of the body is many degrees above the natural temperature. This disease is often accompanied with quickness of breathing and a general uneasiness, with a constant desire to change position, &c.

Treatment.—When the patient is in a cold stage, warm diffusive stimulants, and diaphoretics are indicated, aided by warmth and moisture externally; and in extreme cases apply stimulating liniments to the extremities. For the purpose of promoting perspiration, give the composition drench, made as follows:—

Powdered ginger, . . . 1 ounce.
" cloves, . . $\frac{1}{2}$ "
" cinnamon, . . . $\frac{1}{2}$ "
" capsicum, . . $\frac{1}{4}$ "
Hot water, ... 1 quart.

To be given when sufficiently cool.

If this does not equalize the circulation, apply the foot-bath and manage as directed in article *Mode of Producing Perspiration*. In the first stages of the disease, this simple treatment is all that is necessary. As soon as an equilibrium of the vital fluid is produced, a perfect harmony throughout the system is effected, and disease cannot exist; but when the disease has progressed too far, it is difficult to produce this equilibrium, and to maintain it after it is produced; consequently, it requires a frequent repetition of the medicine. After managing as above directed, if the vital powers be not sufficient to produce a determination to the surface, give the following medicine:—

Pulverized whiteroot, . . 2 ounces.
" skunk cabbage, . 1 "
" lobelia herb, . . 1 "
Boiling water, . . . 4 quarts.

Give one pint once in six hours. If costiveness is one of the symptoms, give

Mandrake root, powdered, . . 1 ounce.
Cream of tartar, . . . 2 "

If this does not operate in six hours, repeat the dose. This may be given in powder, or add one pint of boiling water to the powder, and give the whole as a drench, when sufficiently cool.

SYMPTOMATIC FEVER.

Symptomatic fever is caused by internal inflammation. For example, if the lungs, bowels, or liver is inflamed, the whole system would be thrown into disorder, and a symptomatic fever produced. It is frequently produced by wounds, and especially wounds of the joints.

Symptoms.—Symptomatic fever is not preceded by shivering, nor is it so sudden in its attack as simple fever. Symptomatic fever has many symptoms in common with all other fevers, as loss of appetite, quick pulse, dejected appearance, difficulty of breathing, coldness of the legs and ears, &c. As symptomatic fever is caused by inflammation of different organs, as the lungs, liver, &c., I shall treat of these cases separately under their proper heads.

INFLAMMATION OF THE KIDNEYS.

Horned cattle are often subject to this disorder, and it is too frequently overlooked by the cow or cattle doctors.

Symptoms.—The animal is seized with a trembling fit; it will hump up its back, thrust out its tail, and often attempt to make water, and does frequently in small quantities; and its urine is often high-colored, and sometimes mixed with blood; extremities cold; and the whole frame will be put in motion by sudden starting, caused by pain, and is restless and uneasy.

Causes.—Wounds, bruises, strains across the back, abscesses, swellings, sudden heat or cold, or bad water, is very likely to bring it on.

Treatment.—It is a common custom among physicians and farriers, in inflammation of the kidneys, to give strong diuretic medicines, as balsam of copaiba, oil of juniper, venice turpentine, &c. This is a very dangerous and absurd practice; it is just as absurd to give diuretic medicines in inflammation of the kidneys as it is to give strong or drastic cathartics in inflammation of the mucous membrane of the bowels. The suppression of urine is caused by the inflammation of the kidneys; and any treatment that will increase the inflammation will prevent the secretion. The object should be to relieve internal inflammation, by causing superficial inflammation by means of stimulating liniments, and stimulate the functions of the bowels, skin, &c. With this view I give cathartics and sudorifics. At the commencement of the disease, give

Powdered Mandrake root, . . 1 ounce.
" ginger, . . $\frac{1}{2}$ "
Cream of tartar, . . . 2 "
Gum gamboge, . . $\frac{1}{4}$ "

Mix. Dose, half of the above in a pint of warm water. If this does not operate in six hours, give the remainder. The operation should be assisted by means of injections. After the operation of the physic, give the sudorific drench:—

Fluid extract of veratrum, . . 1 drachm.
Warm water, . . . 1 pint.

If this is not convenient, manage as directed under head of *Mode of Producing Perspiration.* To protect the inflamed parts give large quantities of mucilaginous drinks, such as flaxseed or gum arabic teas.

INFLAMMATION OF THE LIVER.

This is not so common a disease in cattle as some suppose; and it seldom happens to lean cattle, but those that are high fed and kept confined in barns are the most liable.

Symptoms.—Eyes red and watery; tongue furred and yellow; pulse quick and strong; body fuller than it should be for what it eats; constant slavering from its mouth; it also groans much and is short-winded, from the liver being swelled, and pressing hard against the midriff.

Treatment.—The first object should be to stimulate the secretions and excretions, and especially the excretions from the liver and bowels. For this purpose give the following:—

> Pulverized mandrake, . . 2 ounces.
> " blood root, . . 1 "
> " ginger " . . $\frac{1}{2}$ "

Mix, and form into balls about the size of a butternut, by adding common syrup enough to form a ball. Give of these balls sufficient to produce an alterative effect on the bowels, or to keep them sufficiently relaxed. Diuretics are of considerable importance in this disease, and should be freely given. To produce a diuretic effect give the following drench:—

> Balsam copaiba, . . . 2 drachms.
> Spirits nitre, . . . 1 ounce.
> Flaxseed tea, . . . 1 quart.

Mix, and give once a day until it produces its effect on the kidneys.

Vegetables should be freely given in winter; but if it be spring or summer time, herbage in the field will answer better than dry meat. Turning out the animal in the spring will cure many diseases, as it places the animal under natural and healthy conditions, and will be of more service than any system of drugging.

INFLAMMATION OF THE LUNGS.

This is a very common disease among cattle, and probably kills more than all other diseases put together; it frequently makes its appearance in the form of an epidemic, or catarrhal inflammation of the lungs, in which form it is very destructive.

Symptoms.—Among the first symptoms are shivering, with cold extremities, followed by a quick working of the flanks; the membrane of the nose is intensely red, pulse quick and indistinct; the countenance is exceedingly anxious and indicative of suffering.

Treatment.—The animal must be placed under favorable conditions, as regards a comfortable situation, pure air, &c. In this disease, the venous blood is dark-colored and thick, loaded with carbon, and unfit for circulation, and requires oxygen for its decarbonization, this being a very essential principle in the atmosphere. If costiveness is one of the symptoms, take

Powdered aloes,	6 drachms.
" ipecac,	3 "
" lobelia,	2 "
Honey,	4 ounces.
Boiling water,	1 quart.

This should be given when cool, in one dose.

Small doses of ipecac, lobelia, and liquorice should be given occasionally in thin gruel, to relax the tissues, and relaxing injections must be given also.

If this treatment does not produce perspiration and equalize the circulation, apply the foot-bath, and manage as directed in article *Mode of Producing Perspiration.* Counter-irritants are of great importance; by thus inviting the blood to the surface, we relieve the congested state of the lungs, for which purpose use

Tincture of capsicum,	1 pint.
" " ginger,	2 ounces.
Oil of cedar,	2 drachms.

This treatment is very effectual and very little else is needed; for this treatment will equalize the circulation, and remove congestion. The great object should be to maintain the equilibrium of the circulation, and this may be accomplished by occasionally repeating the above-named medicines, and covering the animal with suitable clothing, and flannel bandages applied to the legs.

In chronic inflammation of the lungs we should give something to lubricate the respiratory passages. In such cases I give

Powdered pleurisy root,	.	.	2 ounces.
" liquorice "		.	2 "
" ipecac,	.	.	2 "
" slippery elm,		.	4 "

Mix, and divide the mass into eight powders; to be given in the food morning and evening.

INFLAMMATION OF THE NECK OF THE BLADDER.

This is a very common disease among cows, and especially when a cow is in high condition, and has had a hard time in calving.

Symptoms.—The beast is very uneasy, frequently getting up and lying down, switches its tail, looks back to its hind parts, shifts its position, frequently attempts to make water, which is voided in small quantities and of a high color.

Treatment.—Mucilaginous drinks should be freely given linseed tea is very good, and may be given freely to drink.

Linseed,	. . . .	8 ounces.
Boiling water,	. . .	1 quart.

When cold, give the above to the animal to drink; and give

Cream of tartar,	. . .	1 ounce,
Water,	. . .	1 quart,

as a drench. This should be repeated two or three times a day, as the case may require.

Linseed and bran, made into a mash and given to the cow for food, will serve as medicine and food both.

Warm, diluent clysters are very proper; for this purpose a decoction of marsh-mallow root will answer well; but when you cannot get it make them of linseed tea. The cow should have nothing that is heating; and be sure to keep the bowels relaxed; for this purpose give

Powdered mandrake,	. .	1 ounce.
Cream of tartar,	. .	2 "

Mix, and give the whole at a time, in powder, or as a drench in one pint of water. But if she is kept on proper food, such as linseed, or rye meal, boiled turnips, or carrots, wheat-bran, &c., she will need no physic. Mild diuretics, such as parsley roots, wild carrots, marsh-mallow, made into a decoction by adding boiling water, and when cold given to the cow to drink, are all very good.

INFLAMMATION OF THE STOMACH.

Inflammation of the internal or *villous* coat of the stomach is not a very common disease among cattle, and is generally occasioned by poison or strong medicine that has been given to cure disease, or by the cow's eating poisonous herbs, such as the yew-tree, wild saffron, deadly night-shade, hemlock, foxglove, and other noxious weeds.

Symptoms.—When poisons, or strong medicines incautiously given, are the cause, it will come on suddenly; the pulse will be extremely quick, and so weak that it can scarely be felt; cold extremities; respiration will be disturbed; sometimes there will be a cough, and always a high degree of debility.

Treatment.—The first thing to be done is to give castor or sweet oil; if these cannot be had immediately, give mucilaginous liquids freely, such as decoction of linseed, or gum-arabic, &c.; and, at the same, medicines that are capable of decomposing or destroying the poison, as alkalies, if the poison prove to be either mercurial or arsenical. Clysters are to be injected, composed of strong linseed decoction, or water-gruel; and if purging is one of the symptoms, add one scruple of morphine to each injection. The food should consist of mashes made of bran and linseed meal; for want of the linseed meal, use flaxseed in with the bran. This treatment, given in time, is sure and safe; but when neglected too long, there is no help for the disease.

QUINSY.

This is a very common disease among horned cattle of all kinds.

Symptoms.—In the beginning of this disease the animal will slaver much, thrust out its head, have great difficulty of swallowing—chewing the food and putting it out of the mouth again. On examining, you will find the roots of the tongue much larger than usual; the glands also near the ears are much swollen.

Treatment.—The tincture of belladonna in the human subject is almost specific; and it operates equally well on the animal, and should be given frequently and in small quantities, as a teaspoonful at a time; this should be poured on the root of the tongue, clear, and let it act upon the glands of the throat; at the same time apply stimulating liniments to the throat. The one below will be found very effectual:

Tincture of capsicum,	2 ounces.
Aqua ammonia,	2 "
Oil of turpentine,	1 "
" linseed,	4 "

If the disease has gone too far, the tumor will have to be opened with a knife, or left to break of itself.

When this disease causes much derangement and fever, treat it as directed in *Symptomatic Fever.*

INFLAMMATION OF THE BRAIN; OR, FRENZY.

This is a very common disease among cattle, and prevails sometimes epidemically.

Symptoms.—The animal is nearly blind, and often tosses up its head, till at last a lethargy or sleepiness comes on; the eyes look red; the urine is of a high color, and the dung black and of small quantity; countenance disturbed and frightful, and signs of madness; pulse hard and very slow; trembling of the limbs will be one of the last symptoms.

Treatment.—The first thing to be done in this disease is to equalize the circulation. For this purpose apply the foot-bath, and manage as directed in article *Mode of Producing Perspiration*. Keep the head cold by the constant application of cold water. When the circulation has become equalized, give

Powdered lobelia,	1 ounce.
" mandrake,	1 "
" gum gamboge,	1 "
Cream of tartar,	2 "

Mix, and divide the above into four doses. This may be given in the form of a ball, by adding simple syrup sufficient to form a ball-mass; or add a pint of boiling water to one quarter of the above, and when cool give as a drench. This is a powerful cathartic and nauseating medicine, and will relieve congestion and equalize the circulation. The operation may be assisted by giving stimulating and relaxing injections, composed of

Powdered lobelia seed, . . ¼ ounce.
" capsicum, . . 1 drachm.
" ginger, . . . 2 "
Boiling water, . . 2 quarts.

To be given when blood-warm. This is powerful treatment, but the obstinacy of the disease requires active treatment, and there is no danger of irritating or inflaming the bowels too much with medicine in this disease, as the bowels are very torpid and inactive. This treatment should be repeated as often as necessary, and should be discontinued as soon as the inflammation abates and relaxation is produced. It will then be necessary to give tonics. I then give

Powdered golden seal, . . 4 drachms.
" ginger, . . 2 "

Mix, and give in the feed once a day.

After the disease has abated, be careful of exposing the animal by turning him out, for fear of a second attack.

SPASMODIC AFFECTIONS OF THE MUSCLES, VULGARLY TERMED CROOK.

This is a very common disease in cattle, and is caused by long exposure to cold storms, or by lying on the cold ground too long, and especially when the animal is in poor condition and the blood low.

Treatment.—In the first place manage as directed in article *Mode of Producing Perspiration.* After the circulation has become equalized by this process, and a reaction taken place, give

Powdered lobelia, . . . ½ ounce.
" capsicum, . . ¼ "
Common salt, . . . 1 "

Mix, and add one quart of boiling water. To be given as a drench, when cool.

If the first dose does not succeed in allaying the spasms and relaxing the animal fibers, repeat it in six hours; stimulating liniments should be applied to the parts; and if this treatment does not succeed, give the physic and injections as directed in article *Frenzy, or Inflammation of the Brain.*

Symptoms.—The head is drawn on one side; the animal will look wild, and will thrust its head into any corner; sometimes it is seized with fits of madness at intervals, which are very troublesome; extremities cold, and muscles hard and rigid.

When the animal gets strength, let it have room to stir and turn itself. If the weather is favorable, lay it on the barn floor, with the doors open to give it air; nurse the patient well, and be careful not to expose him too soon, as it would be very likely to bring on a relapse of the disease, which would be very sure to prove fatal.

COUGH, VULGARLY TERMED HOOSE.

A cough is not so common in horned cattle as in the horse, yet they are sometimes attacked with it; as colds or inflammations, either slight or violent, are sometimes received which leave the animal with a cough.

The symptoms are too well known to need any description.

Treatment.—Give the following:—

Powdered elecampane, . . 2 ounces.
" liquorice root, . 2 "
" anise seed, . . 2 "
" honey, . . 4 "

Mix, and divide into six parts; give one every six hours, in linseed tea, or make the medicine into balls, by adding simple syrup sufficient to form a ball-mass, to be divided into eight balls. For food, give boiled linseed or oil cake; this is both food and medicine. All dust should be avoided, and especially dusty or bad hay.

THE YELLOWS, OR JAUNDICE.

This is a frequent disease in cattle, and arises from the liver not performing its function, or becoming torpid and inactive.

Symptoms.—Yellowness of the eyes and nostrils, with costiveness, and a constant itching all over the body; and if the beast have any white hair, it will turn yellow; its dung will be hard and almost the color of burnt clay. Cattle in this disease are not fond of stirring much, it being of a very sluggish nature.

Treatment.—Alteratives, or medicines that operative on the secretions of the liver, should be given until the bowels become sufficiently relaxed, and an alterative effect is produced. For this purpose, use

  Pulverized mandrake, . . 4 ounces.
  "   blood root, . . 2 "
  Extract of dandelion, . . 4 "

Water sufficient to form it into a ball-mass. Divide it into sixteen balls; give three a day until the bowels become relaxed; then give one each evening until the secretions are changed and the animal is better and looks more lively.

This disease leaves the animal in a debilitated condition, and often requires a tonic, for which use

  Golden seal, powdered, . . 3 ounces.
  Ginger root,   " . . 1 "

Mix, and divide into four powders, to be given in the feed three times a day. Good nursing, and a nutritious diet, are necessary to complete the cure.

DROPSY, OR WATER TYMPANY.

This disorder is little understood by many, yet it is a disease of common occurrence in the cow.

Symptoms.—At the beginning of this disease there is nothing to be seen for some time; at length you will perceive the beast to grow fuller every week for some time, and it will not stand nor lie long at a time, for the water will be a burden to it when standing, and still more when lying, as the water presses hard against the midriff, and compresses the lungs and confines their action.

Treatment.—This will consist in stimulating the secretions and excretions; for which purpose make free use of diuretics, or such medicines as will stimulate the kidneys to increased action, for which give freely of

  Fluid extract of Indian hemp (*cannabis*
   apocynum), . . . 4 ounces.
  Tincture of colchicum, . . 3 "

Mix. Dose, one ounce in a pint of water. This should be given once a day. Give the cow to drink a decoction of juniper berries, queen of the meadow, parsley, clivers, or any other herb that possesses diuretic properties.

If this treatment does not succeed, give cathartics, to carry the water off through the excretories of the bowels. The following will be very effectual:—

   Cream of tartar, . . . 4 ounces.
   Gamboge, . . 2 "
   Water, ... 2 quarts.

Dose, one pint once in six hours, until it operates as physic; after this, give a sufficient quantity once or twice a week, to keep the bowels in a relaxed state.

This treatment, if resorted to in time, will cure nine cases out of ten. But it sometimes happens that the disease has progressed too far, and medicine cannot cure, and the patient has to be tapped. I formerly cut them on the fore side, to the udder, but of late years I have done it on the side, opposite to the hip, and about four inches forward of the hip-bone. It is not very material where you cut, if you do not cut a blood vessel of any size, and you will not, in that direction. You need not be much afraid of cutting, as there is little danger in it, for the water lies just within the rim of the belly, and on the outside of the intestines or bowels. It is not like cutting for bloating by eating clover, as then you cut into the great paunch, or first stomach; in this case you have got to cut through where the stomach or paunch grows fast to the side.

DIARRHEA, OR LOOSENESS.

This disease is much better known than the method of cure. It is more difficult to cure this disease in horned cattle than in man, or any other animal. But I never failed to effect a cure with the following treatment, if given in time:—

   Tincture of opium, . . . 2 ounces.
   " kino, . . . 3 "
   " camphor, . . . 1 "
   Essence of peppermint, . . 1 "
   Paregoric, ... 4 "

Mix, and give an ounce once in six hours, until the disease abates; in bad cases, give injections composed of

   Slippery elm flower, . . . 2 ounces.
   Morphine, ... 20 grains.
   Warm water, . . . 1 quart.

The animal should have no water to drink; but should have a liberal allowance of linseed-tea or some other mucilaginous drink. A tea made of the Indian arrow-root has been very effectual in curing the disease. The animal should be kept on dry food, and kept in a warm, comfortable place.

RED WATER, OR BLOODY URINE.

I have seen but few cases of this disease among horned cattle; it is caused by inflammation of the urinary organs, disease of the liver, strains, or bruises.

Symptoms.—The animal stales often, and its water is of a dark, bloody color, and comes in small quantities, and is hot and acrid, and excoriates the neck of the bladder, which causes it to make water little at a time and often; the pulse is hard and full, and general symptoms of fever make their appearance.

Treatment.—Give plenty of mucilaginous drinks, as linseed-tea, and give a decoction of princes' pine three times a day; dose, one quart at a time; and get

Wine of colchicum, . . . 3 ounces.
Tincture of blood root, . . 2 "

Mix, and give one ounce of the above once a day in a pint of flaxseed tea, and relax the bowels with

Cream of tartar, . . . 1 ounce.
Gamboge, ... 1 "
Warm water, ... 1 pint.

To be given as a drench. If this does not operate in six hours, repeat the dose; the operation should be assisted by the use of injections. After the disease abates, and the animal is left weak, give

Unicorn-root, powdered, . . 2 ounces.
Golden seal, ... 1 "
Ginger, ... 2 "

Mix, and give one ounce three times a day in feed. The animal should be housed and kept comfortable; warm and stimulating liniments should be applied to the back, and a nutritious diet, such as linseed and corn meal, carrots, &c., should be given.

FLATULENT COLIC.

I have seen but few cases of the flatulent or wind colic in cattle.

Causes.—Drinking large quantities of cold water when the body has been heated, and the motion of the blood accelerated by violent exercise; bad hay or straw, and particularly chaff; water that is strongly impregnated with iron; snow-water; a sudden change from heat to cold, &c.

Symptoms.—The beast lies down, and quickly gets up again; strikes against its belly with its hind feet, and looks wild; sometimes the pain is so great that the animal will run at anything near it; it looks towards its belly, and throws its head upon its sides, and remains in that position for some time.

Treatment.—At the commencement of this disease, give

Tincture of ginger,	1 ounce.
Paregoric,	4 "
Water or thin gruel,	1 pint.

To be given at a dose.

I have hardly ever found this medicine to fail of giving relief the first time; if it should fail, repeat the dose in two hours, and give, as an injection,

Ginger,	½ ounce.
Common salt,	1 "
Warm water,	1 pint.

FOUL IN THE FOOT.

This is a very common disease among cattle, and especially cattle that run in wet pasture; or when the weather is very wet, they are liable to be affected in any pasture.

Symptoms.—This disease comes on sometimes very suddenly; one day the animal appears well, and the next has a foot much swollen and very lame; its claws are thrown apart, and there will appear a honeycomb-like substance between, and a very disagreeable smell will be noticed.

Treatment.—First remove the animal into a clean, dry pasture, and apply the following ointment:—

Common turpentine, . . 4 ounces.
Rosin, ... 2 "
Beeswax, ... 2 "
Honey, ... 2 "
Hogs' lard, ... 2 "
Verdigris, ... 1 "

Powder the verdigris fine ; melt all the other articles together, and put the verdigris in when they are nearly cold, stirring them well, that it may not settle to the bottom. Apply this salve twice a day, and if there appear any pipes or proud-flesh, eat it down with blue vitriol powdered fine. This treatment I have never known to fail, and I have seen it applied frequently.

ACCOUCHING, OR TO ASSIST A COW IN CALVING.

Natural labor.—By the term *natural labor*, we mean all those labors which occur at the full period of gestation, in which the head presents, and which are completed with ordinary facility, without requiring any artificial assistance from medicine, or from the hand, or instruments. As all natural labor requires no assistance, except in extreme cases of debility, I shall not give natural labor a separate consideration, but shall confine my remarks to

Preternatural labor.—By the term *preternatural labor*, is meant that in which any other part except the head of the animal or calf presents.

Signs of preternatural labor.—It is of the greatest importance to ascertain these cases in the early part of labor; if this is neglected, the opportunity of rendering assistance is lost. Various signs of preternatural labor have been relied on by different practitioners, as peculiar motion of the calf, singular shape of the cow, slow progress of labor, and many others; but the only certain information is to be gained by examination with the hand introduced into the vagina.

In a breech-presentation, or presentation of the lower extremities, most frequently the back of the calf is towards the abdomen of the mother; but sometimes the back of the calf is towards the sacrum of the mother. In all presentations of the inferior extremities, or hinder parts, if the pelvis be of a natural size, and the parts cool and relaxed, nature is fully competent to effect the delivery, and should be left to her unaided powers, as in natural labor. In natural labor, or when the calf is in a natural position, it will have its back to that of the cow, and a fore-leg on each side of its head, and the first thing that can be felt or seen is the calf's fore-feet and nose. When this is the case, there is no fear of the cow not parting with it; but even in this case, she will want a little help sometimes—in cases where the cow is feeble and much debilitated.

In all cases of preternatural labor there is a wrong presentation; and sometimes one foot is right and the other wrong; but be sure to get both feet right before you attempt to draw the calf. Care should be taken that the back of the hand be next the calf-bed or womb, that you may not hurt it with your fingers. If the cow press much against you, raise her hind quarters to give you more room. Sometimes the hind parts of the calf come first, with both the hind-feet down; then let another person put his left arm into the cow to find the tail; then with your right hand fetch the feet up, and then it will come as well as with the head first. If the contraction of the womb and abdominal muscles is great, your arm will be cramped, unless the calf be thrust into her by another person. If you do not let another person introduce his hand into the cow and thrust the calf back, it will be exceedingly difficult to get hold of the hind-legs, when the breach presents itself first with the hind-legs drawn back; if they cannot be got without much difficulty, the calf may be drawn away by the tail. If you require an assistant, let him set his back to your back, and one use the right and the other the left hand, and you will not be in each other's way.

Sometimes the head falls back, and when you put your arm into the cow you cannot find it. It may be under the calf; in which case, turn the calf over, and then try to find it; and you must raise the cow well behind, that you may have more liberty. The task will be a difficult one. Lay the calf in as good a position as you can, and, if the head still continues to fall back, put a hook into the calf's nose, made for the purpose, with the point turned in, that it may catch nothing to do mischief; this hook may be made of large wire, and in the form of a half circle of two inches in diameter; the point must be blunt and smooth. Sometimes the neck of the calf is twisted, but it must be straightened before the calf is drawn.

Twins.—Sometimes there are two calves, one right and the

other wrong. Before attempting to draw either of them, be careful to get hold of two feet of the same calf. There is little danger of a cow that has two calves not doing well; for they are often smaller than when there is only one. It sometimes happens that there are three calves, and only two born, and the third has been left in till it has caused the death of the cow.

When a cow is long in calving, and the membranes or water-bladders break early, the hair of the calf gets dry; in this case, grease it well with lard, or sweet oil, or fresh butter, and it will pass the pelvis much easier.

When the calf is swelled, as in dropsy, the head is so large that the passage is too small for it, and its body is so much swelled that there is no room to draw it nor stir it in the womb; in this case, the calf must be stabbed with a lancet or sharp-pointed knife in many places.

When you are called upon to assist a cow in calving, be cool and deliberate in your proceedings, and never attempt to force nature; let nature do her own work in her own way, and when she fails, assist her. Remember that nature is the only true midwife. Use no force or violence, and never attempt to do anything that you do not understand; if you cannot do any good, do no hurt.

FALLING DOWN OF THE CALF-BED OR WOMB.

This often happens after calving. When this takes place, be careful to wash the calf-bed well, and remove every particle of dirt or anything that adheres to it, and separate the placenta or after-birth, vulgarly termed the cleanings, from the womb, if it can be done without using much force; it may be easily told from the womb by being lighter colored, and of a membranous substance, while the womb is of a dark red, and has the appearance of little glands or rosebuds. If the womb has been down some time before it is discovered (especially in winter), and is cold, be sure to foment it with milk and water to bring it to its natural warmth, before you put it up; for if you put it cold into the cow, it will give her great uneasiness, and cause her to throw it down again, and there will also be danger of its bringing on inflammation.

Some people are so foolish as to put powdered rosin on the

womb to keep it up—it is a sure way to cause the cow to throw it down again; anything that irritates is sure to cause the cow to strain and throw it out again.

When the womb comes down it is something like the lining of a hat when fallen out. When you put it up, have the cow on her feet with her hind feet on the highest ground, oil your hand, put one hand under the lower part of the womb, and when you come to the barren, or what is properly called *labia pudenda*, thrust gently with the back of your hand, and let the lowest part go first. If you cannot get it in in this way, let another person hold the womb in his hands, and then gently work it in by degrees. When you have got it into its place, put your hand into the cow and feel that it is not left doubled, but is rightly placed. Keep her as high behind as you can, that she may not throw it down again. It will be necessary to give an ounce of the tincture of opium, in a pint of hop tea or water gruel, to remove the pain and irritation, that she may not strain and throw it out again.

There are a variety of appliances recommended for keeping up the womb, but all to no purpose. A cruel and barbarous custom prevails among many to this day, which consists in sewing up the barren with a leathern strap; others pierce through the skin of the back with an awl or pricker, and tie, with a sharp cord beneath, to keep the cow from straining in consequence of this arrangement hurting her back. These are barbarous customs, and are the relics of barbarism.

Treatment.—After the womb has been replaced, introduce a common sized sponge, the size of a man's fist, up the vagina, or birth-place. This should be introduced as high up the vagina as can be, without entering the womb; it should pass by the urethra, or passage from the bladder by which the urine passes off, so as not to obstruct it. But what is still better is a *pessary*, which is an instrument to support the womb in prolapsus or falling down, which may be made in the following manner: Take a ball of some soft wood, two inches in diameter, with a hole through the middle a quarter of an inch in size, and coat it over by putting it into melted rosin; enough will adhere to it by dipping once or twice; the coating should be about the eighth of an inch thick. The hole through the middle should not be filled up by the coating, as this enables any morbid matter to escape from the womb. This should be introduced into the vagina, and passed by the urethra or outlet of the bladder, so as not to obstruct the urine in its passage. Before introducing the pessary, attach a piece of ribbon to it, so that it can be removed at pleasure, the end of which should hang out of the vagina two or three inches. This is the most convenient and effectual way of preventing the falling of the womb, or of preventing the cow from casting the weathers, as it is vulgarly termed.

ABORTION, OR LOSING A CALF.

There is a law of sympathy among all animated existence, which is very strong in the cow; for when one cow parts with her calf, there is danger of all the cows in the same yard being thus affected. When you see a cow likely to part with her calf, separate her from the others as soon as convenient, and give the cleansing drink:—

Capsicum, . . . ¼ ounce.
Birthroot, ... 1 "
Warm water, . . . 1 quart.

To be given when cool. This is an infectious disorder, and is conveyed mostly through the sense of smell; and the longer the cleaning is in coming away, the more the air is infected, and the more the contagion spreads. I have known it to spread over a whole village. In olden times, when this trouble prevailed among cows, men would go to the witch-doctors, and give them large sums of money to get rid of the evil; but that was in the days of Salem witchcraft. This, like all other diseases, has been, or may be, traced to a natural cause, as all the operations in nature are governed by cause and effect.

HORN DISTEMPER.

Cause.—From the circumstance that we never knew cattle in good condition affected with this complaint, we attribute it, generally, to poor keeping and hard work.

Symptoms.—The animal appears dull, weak, and languid; taking it by the horn, you will find it cold quite down to the head.

Treatment.—The horn, is generally bored with a gimlet, and found to be entirely hollow and empty; pepper and vinegar is then introduced. This treatment is both cruel and absurd, and originated in ignorance of the disease. What nature has so guarded as to enclose in a tight case of horn, ought not to be laid open to the action of the atmosphere; but to introduce such foreign and acrid or pungent matter into the horn, is still more revolting. It should be treated on more philosophical principles. As the disease originates from a depraved condition, the first thing that should be looked to is to give the animal a liberal allowance of nutritious food, and give alteratives to operate on the secretions and excretions. For this purpose, give

   Powdered mandrake root, . . 4 ounces.
    " blood " . 2 "
    " ginger " . 1 "
    " golden seal " . 1 "

Mix and divide into ten powders; give one each morning for ten mornings, and rasp the horn thin next to the head, and apply spirits of turpentine. Wait ten days after the application of the medicine, and then, if the condition of the animal is not much improved, repeat the medicine as before. This treatment is effectual, if given in time.

CLOVER FEVER, OR BLOATING BY EATING CLOVER.

Treatment.—If there is prospect of immediate or inevitable death, or in less dangerous cases, tapping is a perfectly safe remedy, if performed properly.

Directions for Tapping.—There is a place on the left side, about four inches forward of the hip-bone, and the same distance from the back-bone, or anywhere between the last rib and the hip-bone, where the paunch comes in close contact with the skin, that may be opened without difficulty or danger. In large oxen I have known an orifice so large made, that a man put in his hand and took out the clover, and the animal soon recovered.

In less violent cases, give

   Common salt, . . . 2 ounces.
   Saleratus, . . . 2 "
   Capsicum, or ginger, . . 1 "
   Warm water, . . . 1 quart.

To be given when sufficiently cool. The operation of this medicine should be assisted by stimulating injections frequently thrown into the bowels. If this treatment should fail, resource should be had to tapping.

SCROFULA, OR SCAB.

This disease is not very common, but when it does appear it is often neglected.

Symptoms.—Heat in the skin, attended with itching, and the beast is constantly rubbing itself against anything that is near. In the advanced stages of the complaint, the neck and back of the beast will be nearly covered with scales; in some cases they break out into little ulcers, and run a thin matter which is very offensive. If there are many beasts together, this disorder will spread among them all, if not prevented.

Treatment.—In the first place, give opening and cooling medicines, viz:—

Jalap, ... 4 ounces.
Cream of tartar, . . 2 "

Mix, and divide into three doses, to be given each morning in feed, or as a drench. After giving the first dose, rub with the following:—

Sulphur, ... 8 ounces.
White precipitate, . . 1 "
" hellebore root, powdered, . 4 "
Whale oil, . . . 3 pints.

Mix, and rub the beast carefully all over, and it will cure. Buttermilk is a very good thing to rub them in.

LICE IN CATTLE.

Take white hellebore root, in powder, and rub on dry; or make a decoction:—

Boiling water, . . . 4 quarts,
White hellebore, . . 4 ounces,

and wash the animal all over. This is very good, and hardly ever fails. If this should fail, take

Corrosive sublimate, . . 1 scruple.
Whisky, . . . 1 quart.

Mix, and rub on the animal with a brush. This is a powerful

remedy, but is perfectly harmless; you have but twenty grains of the sublimate to a quart of the whisky; it will not be strong enough to excoriate the skin. This will kill any kind of lice on cattle or horses. When you fail in everything else, try this.

WORM IN THE TAIL, OR TAIL DISTEMPER.

Cause.—It is generally caused by debility and privation of food, or vitiated state of the secretions.

Symptoms.—When you find the animal weak and losing flesh, be sure to examine the tail, and if it be soft towards the lower part instead of having its bony feeling, there will be a cavity filled with water in it, and sometimes a worm.

Treatment.—As this disease originated in depravity, the first thing to be done is to supply the animal with plenty of nutritious food, and a tonic medicine, such as

    Golden seal, . . . 2 ounces.
    Ginger, . . . 1 "

Mix, and divide into three doses; give one each morning. Slit the tail three or four inches, and put in fine salt and bind it up, and the cure is sure.

LOSING THE CUD.

This is all fudge! Some good house-wives will neither rest themselves nor let any of the boys rest until they catch a frog and force it into the creature's mouth; others, more naturally, make a cud out of elder bark and other materials; whilst others have another creature caught whilst in the act of chewing the cud, and steal that to supply the other, thus robbing Peter to pay Paul, and making one patient to cure another according to their theory.

Losing the cud, as it is called, is not a disease, only the effect of disease. Sick cattle frequently remain some time without ruminating or chewing their cud, but are found on recovery to commence the practice without a new cud being thrust into their mouths, for a capital to commence upon.

I once heard an old farmer tell a story of a yoke of oxen of his, that he had been plowing corn with all the afternorn. It was the custom in those times, when they plowed corn with cattle, to muzzle the ox by means of a basket fastened to the nose. It being late at night when the old man turned the cattle out, he forgot to take the baskets off; the result was that in the morning when he found the oxen each basket was filled with cuds. When they laid down to chew their cuds, the baskets were so tight on the nose that the oxen could not chew or masticate them, and kept raising them, and when their mouths got full they would drop into the baskets. And what do you suppose the final result was? Well, the oxen had yet a cud to spare, and a capital to commence operations on.

CHOKED WITH APPLES, POTATOES, &c.

Symptoms.—The creature bloats, stretches, thrusts out its nose, and continues endeavoring to chew, with contortions and agony.

Treatment.—Immediately catch it and brace open its mouth; for this purpose use a common clevis, or an iron ring, as large as you can get into the mouth, or sufficiently large to let the arm through, and examine and see if the substance can be extracted with the hand; if not, and having ascertained it to be an apple or potato, and if it can be felt from the outside through the food-pipe, which may be felt on the left side of the neck, back of the wind-pipe, it may be forced down sometimes by rubbing the finger down on the apple from the outside of the neck; the direction of this pipe may be ascertained by watching a creature that is chewing his cud and seeing the direction it goes in raising and swallowing it. Some recommend forcing it down with a leather-covered whip, but this is a dangerous operation in unskillful hands. A surgeon, being present, can open the gullet and take out the obstruction without difficulty. Some recommend holding a block on one side of the gullet and striking the other side with a wooden mallet, for the purpose of crushing it, but this could not be done without injuring the food-pipe if the apple or potato is very hard.

MURRAIN, OR PLAGUE.

This is a very mortal or fatal disease, and sometimes prevails epidemically.

Symptoms.—This disease presents itself in a variety of shapes or forms, but most generally the animal is stiff, and appears to lose the use of one or both hind legs; shortly after the limb will begin to swell and be exceedingly hot and tender. In a short time the inflammation extends to the body, and mortification ends the suffering of the animal, and the limb will soon turn black—which has given it the name of *Black Leg*.

Treatment.—This should be of the most active and energetic kind, and unless the disease be attacked in its early stages it cannot be controlled by medicine. In the commencement of the disease give

Capsicum, powdered, . . 1 ounce.
Common salt, " . . 4 "
Lobelia herb, " . . . 2 "
Golden seal, . . . 1 "

Mix, and divide into ten doses; to each dose add one pint of boiling water, and when cool pour it down the animal; this should be repeated once in six hours until relief is obtained.

After the inflammatory symptoms have passed give

Golden seal, powdered, . . 2 ounces.
Gum myrrh, " . . 1 "
" camphor, " . . ½ "

Mix, and divide into eight doses; one to be given each day. Apply to the affected leg salt and vinegar, and to the well legs tincture of capsicum. The treatment must be of a stimulating and tonic kind, and energetically applied, if a cure is effected. This disease appears to be contagious, and the sick animal should be separated from the well ones as soon as discovered to be sick.

GARGET, OR CAKE IN THE BAG.

This is a common disease, peculiar to the Cow. It is an obstruction of the lacteals, or absorbent vessels of the lymphatic system, caused by the milk remaining too long in the bag, or by a hurt. Sometimes this disease occurs without any apparent cause.

Symptoms.—The bag is full, but yields not its contents, and is hard and inflamed.

Treatment.—Foment the bag frequently over a decoction of bitter herbs, as wormwood, hops, &c. After fomenting

DISEASES OF DOMESTIC ANIMALS. 207

with the herbs, anoint the bag with the Magnetic Ointment No. 1, under the head of *Ointments*.

There is an herb called poke-weed, which operates on the secretions, and is very good in this case.

STOPPAGE OR SUPPRESSION OF URINE.

This disease arises from various causes, as gravel, inflammation of the kidneys or bladder, but most generally from spasm of the neck of the bladder.

Treatment.—A decoction of parsley root, clivers, dwarf elder, or dandelion, is good; or give

    Spirits of nitre, . . . 1 ounce.
    Oil of juniper, . . . 20 drops.

This is a very convenient and effectual remedy.

HOOVE, OR FOUNDER.

The hoove is a species of founder to which cattle are subject.

Cause.—Eating too much grain, apples, potatoes, &c.

Symptoms.—Swelling of the body, griping, stiffness of the limbs, trembling, groaning, &c., accompanied with a burning fever, which, in bad cases, causes the hair and hoofs to fall off.

Treatment.—As soon as the symptoms are discovered, give

    Soda, ... 4 ounces.
    Water, ... 1 pint.

If soda is not at hand, give saleratus, potash, or some other alkali; if necessary, repeat the dose in smaller quantities. An immediate effect will be seen in the abatement of the symptoms, and in a few hours, commonly, the beast will feed. The philosophy of the cure is this: All the symptoms arise from the stomach being overstrained with food, and thereby rendered incapable of performing its ordinary functions. In consequence of this digestion ceases, and an inflammation of the organ ensues, caused by an excess of acidity, which corrodes its coats and causes the symptoms above mentioned. The substance in the stomach, instead of being digested, forms carbonic acid gas. By giving an alkali, which neutralizes the acid and destroys its corrosive quality, a complete cure is effected. The combination of an alkali with the carbonic acid produces a neutral salt, called the carbonate of the alkali, which is perfectly harmless, and passes off without any detriment. This treatment will be good for the horse as well as cattle in this case.

DISEASES OF SHEEP.

These animals, in their wild state, are subject to few if any diseases. "They live according to the laws of their being; live naturally and healthfully; and, unless they meet a violent death at the hands of man or some other natural enemies, die a natural death." Our domestic animals, as they are generally managed, live under conditions less favorable to health, and sometimes get sick. The fault is generally in the keeper or breeder, and not in the animal, or in the conditions inseparable from its domestic state. With animals, as with man, disease arises from some infringement of the organic laws; but their masters, and not themselves, are responsible for the infringement. When they get sick, in consequence of false conditions under which they are forced to live, man adds insult to injury by forcing his nauseous and poisonous mineral drugs down their reluctant throats. If they recover, in spite of both the disease and remedy, drugs get the credit. When medicine is administered to them, if medicines must be, let it be of the vegetable kingdom—medicines that can be digested and assimilated to their use, and appropriate to their diseases. Animals born of well-developed and perfectly healthy parents, may almost universally be kept in perfect health without drugging. With a sufficient quantity of wholesome food, pure water, protection against storms and cold in winter, complete ventilation and perfect cleanliness in their habitations, and general attention to their comfort and health, there will be little call for medical treatment of any kind; and in rare cases which may occur, we would trust mainly to vegetables as remedial agents, with the co-operation of nature, and healthy conditions as regards diet, air, exercise, and water.

As animals live in accordance with nature's laws, and do not transgress them as frequently as human beings, their diseases are, as a necessary consequence, less numerous and less complicated. And they may be grouped under the heads of *fevers, inflammations, spasms* or *colics, fluxes, eruptions,* and *glandular affections.*

Sheep are so constituted that they do not perspire, consequently have to throw off disease and worn-out particles through other channels or emunctories, as the lungs, kidneys, bowels, feet, &c. The feet of sheep are so organized that they expel or throw off a large quantity of morbid matter from the system, in a healthy state or condition. From these considerations you will see the necessity of keeping sheep on dry land, that their feet may be kept dry and clean. They appear to be governed by instinct when they go a great distance round before they will go through mud or water. Any one might predict the consequence of compelling the sheep to stand in mud and filth constantly.

"The history of the Sheep shows us that it is a denizen of the hills. Its instincts, even in its domestic state, attach it to the upland slopes; and when free to do so, it always seeks the highest grounds, where aromatic plants abound, and the herbage is less succulent than in the valley. The wild Sheep, like the deer, is found to frequent all those places where saline exudations abound, to lick the salt earth."

There is no department in the management of Sheep so little understood as the nature and treatment of their diseases. There are many who, on being informed of the presence of disease in a neighbor's flock, confidently advise the employment of some favorite nostrum, on the empirical supposition that because it cured or was thought to cure one flock, it will cure another. Nothing is taken in account, except that in both cases the afflicted animals are sheep; and it is at once concluded, that what benefited one will benefit another.

The many different circumstances are never thought of in prescribing; oh, no; that would be of no use! Of course it can be of no importance to give a moment's attention to age, sex, situation, &c. These are of trifling moment, and are to be despised by a person armed with a recipe, which some one believes to be like a constable, going through the body and bearing off the intruder.

There are but few diseases among Sheep but what prevail epidemically or are brought on by false habits or conditions in life.

STURDY, OR STAGGERS.

This disease is not of frequent occurrence in the United States, but very common in Great Britain. According to Mr. Youatt, it is nearly confined to sheep from six to twelve months old; after that period, sheep seem to be clear from its attacks through life.

Symptoms.—The sheep cease to ramble with their companions; they are dull, and graze but little; they remain in the most languid and listless condition; separate themseves from the rest of the flock; walk in a peculiar staggering way; seem at times to be unconscious and completely giddy, and sometimes tumble down. In the midst of their grazing, they stop all at once, look wildly around as if they were frightened by some imaginary object, and start away at full speed over the field. In the last stages of the disease, the sheep commences a rotatory motion, and always in one way, and with the head on the same side; he continues to form these concentric circles for an hour at a time, or until he falls; at length he dies emaciated and exhausted.

Treatment.—If the disease is discovered in its early stages, it may be successfully treated by giving powerful diuretics and alteratives, to operate on the secretions; I make use of

 Fluid extract of colchicum, . . 2 ounces.
 "   Indian hemp, . 1 "
 Tincture of blood root, . . $\frac{1}{2}$ "

Mix, and divide into ten doses, one to be given in a half pint of water each day. This is an alterative and diuretic, and will operate on all the secretions and excretions in the body, and will do as much towards cure as any medicine; but the disease is generally beyond the control of medicine, and the most that can be done is in the way of prevention. As the disease is generally caused by running in cold, wet land, as soon as the disease is discovered, remove them into a dry, high pasture, and place them under natural conditions.

APOPLEXY.

This word is taken from the Greek, and means to strike or

strike down. The symptoms of apoplexy cannot be mistaken; they are an abolition of sensation, and cessation of voluntary motion, from suspension of the functions of the cerebrum. This disease is peculiar only to sheep when they are very fat; and their plethoric condition is the inducing cause. The fit rarely occurs if the animal is kept quiet; but brisk exercise and over-fatigue will often bring it on. Sheep, therefore, in high condition, should be driven with great care.

Treatment.—Active cathartics are indispensable in this case; the following should be given:—

    Gum gamboge, . . . $\frac{1}{2}$ ounce.
    Aloes, ... $\frac{1}{4}$ "
    Ginger, ... 1 "

Pulverize and mix; divide into six doses, give one every six hours until you get an effect. Drastic cathartics are the only medicines that will be of any service in this disease, and they frequently fail of effecting a cure.

ŒSTRUS, OR GRUB IN THE HEAD.

Blacklock says: "Much annoyance is caused to the sheep by the presence of animals in the air passages. The gad-fly deposits its eggs on the margin of the nostril, in autumn; these are soon hatched, and the larvæ immediately find their way up the interior of the nose, till they arrive at the frontal sinus, a cavity situated between the layers of the frontal bone, and of considerable size in the sheep. Here they remain until the following spring, when they quit, burrow in the earth for a short season, then become winged insects, and are ready to enter upon the career of torment so aptly gone through by their predecessors."

To prevent the attacks of this mischievous insect, it will be found well, about the beginning of July, and again about the first of August, to assemble the flock, and thoroughly tar the parts adjacent to the nostrils. The effluvium of tar is abhorrent to all winged insects; and hence the philosophy of this treatment.

Few sheep are exempt from grubs in the head, and when the number does not exceed two or three, they will not cause much annoyance. They feed on the mucous secretions of the sinus membrane. But when the number of grubs is larger than common, they produce much irritation, and the sheep will sneeze violently.

Treatment.—Tobacco smoke is the only available remedy, and a very good one, being easily brought in contact with the worms; and, when properly administered, certain in its effects. One person secures the sheep, holding the head in a convenient position, while another, with a lighted tobacco-pipe, places one or two folds of a handkerchief over the opening of the bowl, then passes the stem a good way up the nostrils, applies his mouth to the covered bowl, and blows vigorously through the handkerchief. This operation should be continued for several seconds on each side or in both nostrils.

CATARRH, CALLED CORYZA.

During the winter season this disease is very common among sheep.

Cause.—Exposure to cold storms. Those sheep that are wholly exposed, or whose shelters are imperfectly constructed, are very subject to this disease, which consists in an inflammation of the mucous membrane that lines the nostril and air passages. The chief annoyance is occasioned by an excess of mucus, which clogs the nasal passages, and causes great difficulty of breathing. When a sheep is in this situation, it is believed by some to have a bad cold. In some cases the sheep will die from suffocation.

Preventives—which are always worth more than cure—are good shelter and wholesome food.

Treatment.—First, remove to a warm shelter and comfortable situation; then give expectorant medicines, as:—

Powdered liquorice root, . . 2 ounces.
" ipecac " . . 1 "
" blood " . . 1 "

Mix, and divide into twenty doses; one to be given each day in their feed. This will stimulate the membrane to expel the adhesive mucus, and operate on the secretions and excretions, and produce a healthy action of the glandular system.

HOOVE, OR DISTENTION OF THE STOMACH.

This is a species of founder, and is caused by the sheep

being changed from a poor pasture to a luxuriant one, and gorging itself to an immoderate degree. The gullet is obstructed, and the gases in the paunch cause a remarkable distention, with no passage for escape, except into the chest, down the wind-pipe, sometimes causing suffocation.

Treatment.—As soon as the disease is discovered, give an alkaline drench composed of—

Soda, ... ½ ounce.
Water, ... ½ pint.

This will neutralize the acid, convert it into a neutral salt, which will pass off without injury to the animal. In bad cases, and where there is no chance for giving medicines, the *probang* should be used. This is a flexible rod, made of wood or whalebone—the whalebone rod of an umbrella can be used—having a wooden or ivory ball, about the size of a walnut, securely fastened at the end; which, after being carefully entered, should be forced to the lower extremity of the gullet, which will remove the obstruction, and the gas or wind will pass off.

Prevention.—When it happens that sheep must be put suddenly into good feed, salt them well, or freely, before it takes place; and this should be repeated for several days.

STRETCHES.

This disease is of common occurrence in flocks that are kept exclusively on hay, or other dry food; and is fatal very often, unless an early application of medicine is made.

Symptoms.—The sheep will be frequently changing positions, lying down, and rising up again; frequently stretches, and refuses every kind of food. It is generally caused by costiveness, from being deprived wholly of green food.

Treatment.—The immediate cause being costiveness, reason would dictate cathartic medicines,—as, hogs' lard, or castor oil. When castor oil is given, it may be given in one-ounce doses, repeated at intervals of six hours. The same quantity of lard should be given as of oil, and in the same way. Some use tobacco, with good success; this is a powerful remedy, and should be given only in desperate cases.

Preventive.—Give the flock green food once a day, such as apples, potatoes, or turnips.

DIARRHEA, OR SCOURS.

This is a common and fatal disease in our country, and requires an extensive notice. Youatt remarks:—"If affections of the external coats of the intestines do not frequently occur, inflammation of the inner coats, or mucous membrane, is the very pest of sheep." When the inflammation is confined principally to the mucous membrane of the small intestines, and is not attended by much fever, it is termed *diarrhea;* when there is inflammation of the large intestines, attended by fever, and considerable discharge of mucus, and some blood, it is called *dysentery.* The diarrhea of lambs is a dreadful disease. If they are incautiously exposed to cold, or the mother's milk is not good, or if they are suckled by a foster-mother that had weaned too long before, a violent purging will suddenly come on, and destroy them.

When the lamb begins to crop the grass at the side of the mother, he is liable to occasional disturbance of the bowels; but as he gains strength the danger is over. At weaning-time, care should be taken of him. The farmer ought not to be in haste to stop every little looseness of the bowels, for it is often a sanative process of nature; but, if it continues longer than twenty-four hours, and much mucus is discharged, and the appetite of the animal is failing, it will be necessary to attend to the case.

Treatment.—Make use of the following remedy:—

Tincture of kino, . . 1 ounce.
" " opium, . . . $\frac{1}{2}$ "
" " camphor, . . $\frac{1}{4}$ "
Paregoric, ... 2 "

Mix, and give to a lamb a teaspoonful once in six hours, until relief is obtained; for a full grown sheep, give a tablespoonful at a dose.

Give the animal dry food, and plenty of mucilaginous drinks, as arrow-root and linseed teas, cold.

Diarrhea often attacks the full grown sheep, and is frequently fatal, and especially when it has degenerated into dysentery. It is common in the spring, and particularly when the new grass begins to start. If the looseness continues too long, sheep should be removed to shorter and dryer pasture, and hay should be offered to them; if the looseness does not abate, then adopt the treatment above recommended. Sheep should not be changed too suddenly from dry to green food.

Diarrhea can be easily arrested by mixing a small quantity of gum kino, or alum, in wheat bran and salts, and feed dry for a day or two. A decoction of hemlock bark has been used with good success. But, when all these remedies fail, use the tinctures as above directed.

ACUTE DROPSY, OR RED WATER.

Dropsy is a common disease in American flocks. Sheep frequently are destroyed by this disease, without showing any symptoms of illness whatever. The shepherd leaves his flock at night, after a minute examination; and, on his return in the morning, a sheep will be found dead, lying nearly in the usual posture, the legs bent under him, and the head protruded. Apparently there has not been any severe struggle. On examination after death, the belly will be found to contain a large quantity of bloody fluid.

Cause.—A change of pasture, and especially from a dry to a cold and damp one, cold, white frosts, &c. The belly, coming most in contact with the damp and cold ground, is the first affected; the peritoneal coat of the intestines becomes chilled; reaction and inflammation soon follow; its natural function, the secretion of a fluid to lubricate the cavity of the belly, is morbidly and strangely increased; the fluid accumulates, and it is red and bloody from the rupture of small vessels of the peritoneum distended by inflammation. The inflammation pursues its course with almost incredible rapidity, and the animal soon dies.

Treatment.—There is no mode of treatment that will be of any use as long as the sheep remains in this unhealthy situation. The first thing to be done in this case is to remove the sheep to a dry and warm place, and give the following:—

Castor oil, ... ½ pint.

After the operation of the physic, give an infusion of spice-bush tea for a common drink, and be careful not to expose the animal too soon to the weather

DYSENTERY.

Dysentery is essentially inflammation of the large intestines.

Cause.—It is often the result of neglected or obstinate diarrhea, although distinct from it; and sometimes the consequence of unwholesome food, or being pastured on wet or ill-drained meadows, and of being half-starved even there. Fever is a constant attendant on it in its early stages, and wasting and rapid debility soon follows.

The discharge of dysentery is different from that of diarrhea; it is thinner and more adhesive, and often mixed with mucus and blood, which causes it to cling to the tail and thighs, and there it accumulates, layer after layer, a nuisance to the animal, to the owner a warning of much danger, and that near at hand. When this kind of dysentery attacks the animal, it occasionally carries off its victim in a few days, but frequently he lives five or six weeks.

Treatment.—In the first place, give the following mixture:—

Rhubarb,	1 ounce.
Soda,	1 "
Peppermint herb,	¼ "
Boiling water,	1 pint.

When cool, strain, and add half pound of loaf sugar. Dose, a tablespoonful once in three hours. This mixture will remove morbid matter from the bowels, and neutralize any acid that may be in the stomach or bowels. After this medicine has been given for twelve hours, give

Tincture opium,	¼ ounce.
" kino,	¼ "
" camphor,	½ "
Paregoric,	2 "

Mix, and give a teaspoonful once in three or six hours, as the case may require. Give plenty of mucilaginous drinks, as slippery elm, flaxseed tea, &c.

The sheep must not be turned on the same pasture from which he was taken; let him be turned out on high and dry land.

POISON FROM LAUREL.

Sheep and calves will often, in the winter and spring of the year, eat greedily of the low laurel.

Symptoms.—The animal appears dull and stupid, swells a little, and is constantly gulping up a greenish fluid, which is swallowed down; a part will trickle out of its mouth, and discolor its lips. The laurel brings on a fermentation in the stomach, and nature endeavors to throw off the poisonous herb by retching and vomiting.

Treatment.—In this complaint, give sweet milk, or what is better—

Soda, ... ½ ounce.
Soft water, ... ½ pint.

Mix, and give it as a drench. This, if given in time, is a sure cure.

THE ROT.

This disease is classed among those of the liver, because when the animal dies of this malady, the mischief is found in that organ.

Happily for the American farmers, this destructive malady is, comparatively, of unfrequent occurrence in their flocks; but in Great Britain, on the authority of Mr. Youatt, more than a million of sheep and lambs die every year by this disease. "In the winter of 1830–31, this number was more than double; and had the pestilence committed the same ravages throughout the kingdom which it did in a few of the middle, eastern, and southern counties, the breed of sheep would have been, in a manner, extirpated." Many of the farmers lost their entire flocks.

This disease is not peculiar to England. Many sheep are destroyed by it in Germany. In the north of France they are frequently swept away by it; and in the winter of 1809, the ravages were terrific throughout the kingdom. It has prevailed at some periods nearly over all Europe, as far north as Norway.

Symptoms.—The early symptoms of this disease are exceedingly obscure; this is much to be deplored, because in the first stage of it alone does it admit of a cure. The animal is dull, lagging behind his companions; he does not feed so well as usual; if the wool is parted about the brisket, the skin will have a pale yellow hue. The eye of the sheep, in the begining of the rot, can never be mistaken—it is injected, but pale; the small veins at the corner of the eye are tinged, but they are filled with a yellow serous fluid, and not with blood. Farmers pay great attention to this in their examination or purchase of sheep. If the fleshy excrescence about the eye is red, they have a proof which never fails them that the animal is healthy. There is no loss of condition, but quite the contrary, for the sheep in the early stages of the disease has a great propensity to fatten. Mr. Bakewell was aware of this, for he used to overflow certain of his pastures, and when the water had drained off, turned those of his sheep there which he wanted to prepare for market. They speedily became rotted, and in the early stage of the rot they accumulated flesh and fat with wonderful rapidity. By this maneuver he used to gain five or six weeks on his neighbors.

" As the disease becomes confirmed, the yellow tinge begins to spread; the muzzle and tongue are stained; the animal is more dull and dispirited; his false condition rapidly disappears; the membrane of the nose becomes livid; the tongue gradually assumes the same character; the eyes are dull, and their vessels charged with a yellow-brown fluid. The breath now becomes fetid, the bowels variable—sometimes costive, and at other times loose to a degree that defies the power of medicine. The skin often becomes spotted with yellow, or black; the emaciation is more and more rapid; the general fever increases; the vessels of the eye are more distended and red; the skin becomes loose and flabby, and if pressed upon, a peculiar crackling sound is heard; the wool comes off when pulled with the slightest force; the appetite entirely fails; the belly begins to enlarge, and, on pressure, fluid is recognized within it, and hence one of its names, ' the hydropic,' or dropsical rot. The animal is weak in every limb; a violent purging is now very frequently present; the sheep wastes away to a mere skeleton, and at length he dies; the duration of the disease being from two to six months."

" When a rotted sheep is examined after death, the whole cellular tissue is found to be infiltrated, and a yellow, serous fluid everywhere follows the knife. The muscles are soft and flabby, and have the appearance of being macerated. The kidneys are pale, flaccid, and infiltrated. The belly is filled with water, or purulent matter; the peritoneum is everywhere thickened, and the bowels adhere together by means of an unnatural growth. The heart is enlarged and softened, and the lungs are filled with tubercles. The principal alteration of structure is in the liver. It is pale, livid, and broken down with the slightest pressure; and on being boiled, it will almost dissolve away.

Sometimes the liver is curiously spotted; in some cases it is speckled like the back of a toad. Here is the decided seat of the disease, and it is here that the nature of the disease may be learned. *It is inflammation of the liver.* In consequence of this, secretion from the liver is increased, at first scarcely vitiated, and the digestive powers are rendered more energetic; but soon the bile flows so abundantly that it is taken into the system, and the eye, the brisket, the mouth, become yellow. As the disease proceeds, the liver becomes disorganized, and its secretions more vitiated, and even poisonous; and then follows a total derangement of the digestive powers."

Cause.—The rot in sheep is evidently connected with the soil or state of the pasture. It is confined to wet seasons, or to the feeding on grounds moist and marshy at all seasons. It has reference to the evaporation of water, and to the presence and decomposition of vegetable matter. It is rarely or never seen on dry or sandy soils and in dry seasons. In the same farm there may be certain fields on which no sheep can be turned with impunity, and others that seldom or never give the rot.

Some seasons are far more favorable to the development of the rot than others, and there is no doubt as to the character of these seasons. After a rainy summer, or a moist autumn, or during a wet winter, the rot destroys like a pestilence. A return and continuance of dry weather materially arrests its murderous progress. It is therefore sufficiently plain that the rot depends upon, or is caused by, the existence of moisture. A rainy season, and a tenacious soil, are frequently or inevitably sources of it.

Treatment.—In the early stages of this disease, alteratives should be given to act upon the secretions of the liver, for which use

Soda,	1 ounce.
Powdered mandrake root,	$\frac{1}{2}$ "
" bloodroot,	$\frac{1}{2}$ "
" dandelion root,	1 "

Mix, and divide into twelve doses; give once in six hours until you get an effect upon the bowels; then give a powder once a day for eight or ten days; then give, as a tonic,

Golden seal, . . 2 ounces.
Ginger, . . . 1 "

Mix, and divide into ten doses,—one to be given each day in feed or ball.

To this should be added—a simple and cheap medicine—*common salt.* The valuable properties of salt in promoting the condition, and relieving and preventing many of the diseases of animals, is beginning to be appreciated. In the first place it is a purgative, inferior to few, when given in full doses; and it is a tonic as well as a purgative. Its first powers are exerted on the digestive organs—on the stomach and intestines—augmenting the secretions and quickening the energies of each. It is the stimulus which nature herself points out; for, in moderate quantities and mingled with food, men and beasts are fond of it. The sheep, having a little recovered from the disease, should still continue on the best and dryest pasture on the farm, and should always have salt within their reach,—rock salt is the best.

INFLAMMATION OF THE LUNGS.

This is a frequent disease among sheep.

Cause.—Cold and wet pasture; chills after hard driving; washing before shearing, when the weather is too chilly and wet, and other circumstances of a similar description.

Symptoms.—Its first indications are those of fever, hard and quick pulse, slight heaving of the flanks, and a frequent and painful cough. The disease soon assumes a more aggravated form, but further description would be useless. It is sufficient for the farmer to know the first symptoms of the malady, and then pursue that course of treatment which is proved to be the best.

Treatment.—As soon as this disease is discovered, give the following medicine:—

Lobelia root, . . . 1 drachm.
Liquorice root, . . 4 "
Bloodroot, . . 2 "
Mandrake root, . . 2 "
Boiling water, . . . 1 quart.

When cool, give one gill every six hours until it operates on the bowels. Then give the following powder:—

Pulverized liquorice root, . 2 ounces.
" ipecac " . 2 drachms.

Mix, and give a teaspoonful three times a day, in wheat bran, as long as the cough remains; and secure for the sheep a comfortable place, free from all exposure to vicissitudes of the weather. Avoid all food of an irritating nature. Mashes of wheat bran will be found excellent, with a little salt occasionally.

DROPSY.

This disease is caused by long exposure to cold and wet weather. Tapping is condemned by experience and observation.

The best plan of treatment is to give freely of diuretic medicines to operate upon the kidneys, and cathartics to operate upon the bowels. For this purpose give

Tincture of Indian hemp (*apocynum cannabinum*), . . . 1 ounce.
Tincture of colchicum, . . 2 "

Mix, and give a teaspoonful three times a day in a half pint of water. Here you have a medicine that will act upon the kidneys and bowels combined, and is a sure cure for dropsy in man or beast, if given in time and in proper quantities. After the bloat goes down, care should be taken that the disease does not return by exposure to wet and cold situations.

ABORTION, OR LOSING THE LAMBS.

This disease is not so common as in cows, but sometimes occurs very extensively in flocks of sheep. Ewes are liable to it through every stage of pregnancy, but generally it occurs when they are about half gone. The causes are various: sudden fright, jumping over ditches, and being worried by dogs, and too free use of salt; but the most frequent cause is the unlimited use of turnips and succulent food.

Symptoms.—The first manifestations are dullness and refusal to feed; the ewe will be seen moping at a corner of the lot, and will be heard to bleat more than usual. To these symptoms succeed restlessness, and often trembling, with slight labor-pains, and in the course of twelve hours abortion will take place. Sometimes the parts will be so relaxed that the *uterus*, or womb, will become inverted, and the expulsion of the placenta, or after-birth, will precede that of the fetus or lamb.

As soon as the condition of the ewe is discovered, place her in a dry situation, and give the following medicine, with some nourishing gruel.

Tincture of opium, . . 1 drachm.
" camphor, . $\frac{1}{2}$ "

Mix, and give in a pint of gruel; if the symptoms do not abate, repeat the dose the next day.

INVERSION OF THE UTERUS, OR FALLING OF THE WOMB.

This may take place in the ewe at any period, from sudden severe exertion, or straining hard, but is most frequent immediately or very shortly after parturition, or lambing. In this case, it arises from the violent spasmodic action of the womb, which turns inside out, and protrudes out of the body.

Treatment.—No time should be lost in replacing. The ewe must be placed on her back, with her hind-feet elevated; and the hands being lubricated with oil or lard, the womb should be gently forced back into its natural situation. Previous to replacing the womb, it should be cleansed from all dirt, and be brought to the natural temperature of the body by fomenting with milk and water at blood heat; if put up cold, as it would become in cold weather, the ewe would immediately reject it, and throw it down again. After it is replaced, give

Tincture of opium, . . 20 drops.
Gruel, . . . 1 pint.

Mix, and give at a dose, and keep the ewe perfectly quiet.

GARGET.

This is caused sometimes by constitutional derangement, but generally by the death of the lamb. The milk in the udder or bag becomes coagulated.

The udder should be fomented with water as hot as it can be borne. The fomentation, if necessary, should be repeated, and then camphor ointment rubbed on twice a day. If this is not to be had, use bitter-sweet ointment. If the swelling continues, and matter forms, it should at once be opened, and let the pus pass out.

THE SCAB, OR ITCH.

This disease of the skin is exceedingly common among sheep almost all over the civilized world. According to Mr. Youatt, there are several varieties of it.

In the first stages of the disease, there is no appearance whatever of a cutaneous disease. The ordinary scab in sheep is much akin to the mange in other animals. It is most common in the spring and early part of summer. It has been produced by a variety of causes, such as bad keeping, and exposure to cold, and hot wet weather. The prevailing cause, however, is contagion.

Symptoms.—The sheep is restless—scratching and nibbling itself, and tearing off the wool. When closely examined, the skin will be found to be red and rough. Numerous pustules have broken and run together, and form small or large patches of crust or *scab;* hence the name of the disease. The general health of the animal is affected according to the extent and virulence of the disease or eruption; sometimes he pines away and dies, exhausted by continued irritation and suffering. It is a most contagious disease; if it is once introduced into a flock, the farmer may be assured that, unless the diseased sheep are immediately removed, the whole of his flock will become infected.

"After it was found that the itch in the human race was caused by an insect, a species of *acarus*, it was supposed that similar cutaneous diseases in animals might arise from the same cause. M. Walz, a German, was the first to establish this point, and fully investigate its character; and numerous subsequent examinations have proved the correctness of his opinions. He found that the scab, like itch, mange, &c., is caused by animalculæ; that the irritation caused by burrowing in the skin forms the pustule; and that when this breaks, the acarus leaves his habitation and travels to another part of the skin, and thus extends the disease. When any of these acari are placed on the wool of a sound animal, they quickly travel to its roots, where the place of burying themselves is shown by a minute red point. About the sixteenth day the pimple or pustule breaks, and if the acarus is a female it appears with a multitude of young. These immediately set to work on the skin, bury themselves and propagate, until the poor animal is irritated to death, or becomes incrusted with scab. M. Walz satisfactorily traced the parasite through all its changes, and by experiment discovered its mode of action, and mode of infection. He found that when the male acarus was placed on a sheep, it burrowed; the pustule was formed; but the itching and scab soon disappeared, without the employment of any remedy. Such was not the case where the female acarus was placed on the sound skin; as, with the breaking of the pustule, from eight to fifteen little ones made their appearance. M. Walz found that the young acarus, kept in a dry place, dried and crumbled to dust; but when old, it would retain its life through the whole winter; thus proving the necessity of not relying on the season for their destruction, but on preparations of active medicine, when the disease shows itself. Of the origin of these insects we of course can know nothing; it is enough that we are certain when they make their appearance they can be met and destroyed.

Treatment.—The wool should be sheared off about the pustules, the scab should then be removed with a knife or comb, and apply the common precipitate itch ointment, or apply a decoction of tobacco, or a decoction of hellebore, mixed with vinegar, sulphur, and spirits of turpentine.

Preventive.—As an ounce of prevention is worth more than a pound of cure, the best recipe is in the shape of prevention, which consists in warm and dry shelter for the flock during the winter, and wholesome and nutritious food the year round. A poor sheep will always be the first to suffer from this loathsome disease.

ERYSIPELAS.

This is not a very common disease in the Sheep, but when it does make its appearance it may be known by the following

Symptoms.—The first appearance of the disease is that of a red, inflammatory thickening of the skin, breaking out in a fine eruption, frequently watery, attended with fever and heat. It attacks, most generally, those sheep which are in the best condition, and has sometimes proved very fatal, it being a disease which does not run long before it kills the animal.

Examination after death generally shows an inflammation of the stomach, kidneys, intestines, bladder, &c.

Cause.—Feeding on too succulent food, and running in too low pasture land.

Treatment.—Change of diet, and cooling purgative medicines freely administered. The following prescription has been used with good success:—

   Powdered mandrake root, . . 1 ounce.
   Cream of tartar, . . 2 "

Mix, and divide into six powders; give one every six hours until you get an effect; then give one each morning for some time. After the effect of the purgative medicine, give

   Golden seal, . . . 1 ounce.
   Ginger root, ... $\tfrac{1}{2}$ "

Powder and mix, and divide into six powders; give one each day.

JOHNSWORT SCAB.

The weed called *Johnswort* will cause an irritation of the skin, often over the whole body and legs of the sheep; but generally it is found to be confined to the neighborhood of the mouth.

If eaten in too large quantities, it will produce violent inflammation of the bowels, and is frequently fatal to lambs, and sometimes to adults. Its effects are very singular; the sheep will sometimes run in a circle, with all the precision of a circus-horse; this will be continued until they will fall from exhaustion.

Treatment.—If there are symptoms of the disease as above described, administer

   Sweet oil, . . . $\tfrac{1}{2}$ pint.

If, after an hour, the symptoms do not abate, give

   Soda, ... 2 drachms.
   Water, . . . 1 pint.

Give this as a drench, and anoint the irritated parts with hogs' lard and sulphur. Remove the flock to pasture free from the weed, and salt freely. It is said that salt, given often to sheep, is an effectual guard against the poisonous properties of the weed.

PELT-ROT.

This disease of the skin is caused by exposure during the winter, and low condition—the latter principally.

Symptoms.—This disease may be known by a falling off of

the fleece in the spring of the year, without any eruption of the skin.

Preventive.—Good shelter and good keep. This will keep the wool-fluids healthy and abundant, and there will be no danger of an attack from this disease.

SORE MOUTH.

It is generally supposed that this disease is caused by sheep eating noxious weeds. A correspondent of the "Cultivator" thus speaks concerning it: "It generally commences in one corner of the mouth and spreads over both lips, and the lips swell to the thickness of a man's hand. My flock consisted of about three hundred, and in the space of three weeks about forty died of the distemper, and not one had recovered. By this time, at least one half of the remainder of the flock were attacked. It occured to me that tar would be as likely as anything to give relief. I accordingly had my sheep all brought together; and filled their mouths and daubed on their lips all that could be made to stick; and, to my surprise, it effected an immediate cure. I lost but two or three after this, and these were nearly dead when I made the application. In a few days every sheep was well." Hogs' lard and sulphur will also cure the disease.

MAGGOTS.

In the spring of the year, sheep are subject to the scours, or diarrhea, which causes an accumulation of filth about the tail, and attracts the maggot-fly. If maggots are at work about the tail, the sheep will be seen biting, and rubbing about against the fences and every object that presents itself.

Treatment.—Dislodge the worms with the knife, and apply spirits of turpentine. If they have penetrated far into the skin, hold the sheep in such a position as to retain the liquid for a minute or more in the affected part. By so doing the maggot will crawl out and perish instantly. Sheep cannot be too closely watched before they are shorn, otherwise some will be destroyed from the above cause. The maggot and the tick can be destroyed by tobacco decoction.

FOOT-ROT.

This is a common scourge of sheep, throughout all parts of the United States, and requires an extensive notice as regards the cause and most approved mode of treatment. I have never had any personal observation of this loathsome malady. For this reason I am compelled to rely on the scientific accounts of Youatt, and Professor Dick, of Edinburgh, as to the cause, and to intelligent sheep-breeders of our own country for its treatment. Mr. Youatt says,—

"Foot-rot is a disease, at first, and usually throughout its whole course, confined to the foot.

"*Symptoms.*—The first indication of foot-rot is a certain degree of lameness in the animal. If he is caught and examined, the foot will be found hot and tender, the horn of the hoof soft as usual, and there will be an enlargement about the coronet, and slight separation of the hoof from it, with portions of horn torn away, and ulcers formed below, and a discharge of thin, fetid matter. The ulcers if neglected, continue to increase; and throw out fungous granulations; they separate the hoof more and more from the parts beneath, until at length it drops off.

"*Cause.*—All this is the consequence of soft and marshy pasture. The mountain or down sheep—the sheep in whose walk there is no poachy ground—if he is not actually exposed to infection by means of virus, knows nothing of this disease; it is by the yielding soil of low country that all the mischief is done."

The following is from the pen of Professor Dick:—

"The foot presents a structure and arrangement of parts well adapted to the natural habits of the animal. It is divided into two digits, or toes, which are shod with hoofs composed of different parts, similar in many respects to the hoof of the horse. Each hoof is principally composed of the crust or wall, and the sole. The crust, extending along the outside of the foot, round the toe, and turning inwards, is continued about half way back between each toe on the inside. The sole fills the space on the inferior surface of the hoof between these parts of the crust, and being continued backwards becomes soft as it proceeds, assuming somewhat the structure of the frog in the foot of the horse, and performing, at the same time, analogous functions. The whole hoof, too, is secreted from the vascular tissue underneath."

Now this diversity of structure is for particular purposes. The crust, like that of the horse, being harder and tougher than the sole, keeps up a sharp edge on the outer margin, and is intended to resist the wear and tear to which the foot of the animal is naturally exposed. The crust, therefore, grows unrestrained, until it either leaps over the sole, and serves to retain and accumulate the earth and filth, or is broken off in detached parts—in some cases exposing the quick, or opening new pores, into which particles of earth or sand force their way, until, reaching the quick, an inflammation is set up, which in its progress alters or destroys the whole foot. The finest pastures and richest lawns are particularly liable to this disease, and so are soft, marshy, and luxuriant meadows. It exists, to a greater or less extent, in every situation that has a tendency to increase the growth of the hoofs without wearing them away.

Sheep that are brought from an upland range of pasture are more particularly subject to it. This is very easily accounted for. By means of the exercise which the animal was compelled to take on account of the scantier production of the upland pasture, and also in consequence of the greater hardness of the ground, the hoof was worn down as fast as it grew; but on its new and moist habitation, the hoofs not only continued to grow, but the rapidity of that growth was much increased, while the salutary friction which kept the extension of the foot within bounds was altogether removed. When the nails of the fingers or toes of the human being exceed their proper length, they give him so much uneasiness as to induce him to pare them, or if he neglects this operation they break. He can pare them after they have been broken, and the inconvenience soon ceases. When, however, the hoof of the sheep exceeds its natural length and thickness, that animal has no power to pare it off, but its length continues to wound, irritate, and induce to separate, by the introduction of foreign and annoying matters into it. The different parts of the hoof, likewise, deprived of their natural wear, grow out of their proper proportions. The crust, especially, grows too long; and the overgrown parts either break off in irregular rents, or by overshooting the sole, allow particles of sand and dirt to enter into the hoof. These particles soon reach the quick, and set up inflammation, as already described, and followed by all its destructive effects.

The ulceration of the foot-rot will not long exist without the additional annoyance of the fly. Maggots will soon multiply on every part of the surface, and burrow in every direction. To this, as may be readily supposed, will be added a great deal of constitutional disturbance. A degree of inflammatory fever is produced. The animal for a while shifts about on its knees; but at length the powers of nature fail, and it dies from irritation and want.

Treatment.—The foot must be carefully examined, and every portion of loose and detached horn pared off, even though the greater part, or almost the whole of the hoof, may be taken away. The horn once separated from the parts beneath, will never again unite with them, but become a foreign body, and a source of inflammation and fungous sproutings. This, then, is the first, and fundamental thing—every portion of horn that is in the slightest degree separated from the parts beneath, must be cut away.

If there are any fungous granulations, they must be cut down with the knife, unless they are exceedingly minute, and then the caustic about to be mentioned will destroy them. The whole hoof must be thoroughly cleaned, although it may occupy much time, and inflict considerable pain on the animal. The foot should then be washed with a solution of chloride of lime:—

Chloride of lime,	4 ounces.
Nitre,	1 quart.

This will remove the fetor, and tendency to sloughing and mortification, which are frequent attendants on the foot-rot. Then apply the following:—

White vitriol,	1 ounce.
Sugar of lead,	1 "
Crocus martis,	¼ "
Water,	1 quart.

If the foot has been in a manner stripped of its horn, and especially if a considerable portion of the sole has been removed, it should be wrapped in a little clean tow, and bound tightly down with tape; the sheep should be removed to a straw-yard, or some enclosed place, or to a dry pasture. This is absolutely necessary; for if, when the sheep is again turned out, the foot is exposed to the original cause of disease, the evil will return under an aggravated form.

The sooner the bandage can be removed, and the sheep turned into some upland pasture, the better it will be for the foot, and the health of the animal generally.

The sheep that has had the disease should not be suffered to join his companions while there is the slightest discharge from any part of the hoof, for the disease is highly contagious.

The following recipes for the foot-rot have been used successfully by American sheep-breeders:—

Blue vitriol, . . . 4 ounces.
Verdigris, . . . 2 "
Urine, ... 1 pint.

Another Recipe.

Spirits turpentine, . . . 4 ounces.
Tar, ... 4 "
Verdigris, ... 3 "

This is an excellent ointment to apply to the foot that has been much cut away, before the application of the tow.

The following remarks will show that the foot-rot is contagious beyond all question. A farmer having been sorely plagued with the disease in his flock, fequently renewing itself after repeated cures, resolved to slaughter the whole, which was accordingly done. Several months afterwards, he purchased another flock, which were never known to have been infected, nor was the disease known in the vicinity where the purchase was made; and lo! in less than one month after they were brought to their new home, the sheep became diseased like the previous flock.

FOULS.

The fouls is another variety of the foot-rot, and is produced by stubble-land. The rubbing of the stubble causes a friction between the clefts of the hoofs. The gland in that situation swells, becomes enlarged, and suppurates. This complaint is, however, more readily remedied than the other, and does not cause nearly so much suffering to the sheep.

Treatment.—The application of tar and turpentine will cure the disease, if removed from the cause that produced it. It is not considered contagious.

DISEASES OF THE SWINE.

THE Swine, in its wild and natural state, is a bold and powerful creature, and becomes more fierce and indocile by age. The wild boar, which was undoubtedly the progenitor of all the European varieties, and also of the Chinese breed, was formerly a native of the British islands, and very common in the forests until the time of the civil wars in England.

In the wild state, the Hog has been known to live more than thirty years; but when domesticated, he is usually slaughtered for bacon before he is two years old.

When the Wild Hog is tamed, it undergoes many changes in its formation and habits; and all the varieties of breed have been caused by climate, and conditions under which the animal lives. From the form of his teeth, he is chiefly herbivorous in his habits, and delights in roots, which his acute sense of smell enables him to discover beneath the surface. He also feeds upon animal substances, such as worms and larvæ, which he grubs up from the ground, the eggs of birds, small reptiles, the young of animals, and occasionally carrion; he even attacks venomous snakes with impunity.

From the various allusions to the Hog in the writings of the ancient Greeks and Romans, it is plain that the animal existed in their day. Varro states that the Gauls produced the finest swine that were brought into Italy; and according to Strabo, in the reign of Augustus, they supplied Rome and nearly all Italy with gammons. This nation, and the Spaniards, appear to have kept large droves of swine.

On the other hand, swine's flesh has been held in utter abhorrence by the Jews since the time of Moses, in whose laws they were forbidden to make use of it as food. The Egyptians, also, and the followers of Mahomet, have religiously abstained from it. Paxton, in his "Illustration of Scripture," says: "The Hog was justly classed by the Jews among the vilest animals in the scale of animated nature; and it cannot be doubted that his keeper shared in the contempt and abhorrence which he had excited. The prodigal son, in the parable, had spent his all in riotous living, and was ready to perish through want, before he submitted to the humiliating employment of feeding swine."

"Swine," Herodotus says, "are accounted such impure beasts by the Egyptians, that if a man touches one, even by accident, he presently hastens to the river, and, in all his clothes, plunges into the water. For this reason, swine-herds alone of the Egyptians are not allowed to enter any of their temples; neither will any one give his daughter in marriage to one of the profession, nor take a wife born of such parents, so that they are necessitated to intermarry among themselves."

The Brahminical tribes of India share with the Jews, Mohammedans, and Egyptians, this aversion to the Hog. The modern Copts, descendants of the ancient Egyptians, rear no swine; and the Jews of the present day abstain from their flesh as food.

The flesh of swine, and especially in hot climates, is not good, and should not be taken as food; and Mr. Sonnini remarks that in "Egypt, Syria, and even the southern parts of Greece, this meat, though very white and delicate, is so far from being firm, and is so overcharged with fat, that it disagrees with the strongest stomach. It is decidedly unwholesome, and this will account for its proscription by the legislatures of the East. Such abstinence was indispensable to health under the burning suns of Arabia and Egypt." How is it under the burning suns of Carolina and Georgia, and even in our own climate, and especially in the summer season? It may be necessary, in the coldest part of the season, in some constitutions, in our climate, to take animal food to assist in keeping up animal heat, it being of a more positive nature than vegetables. But, where it is necessary, let them eat animal food that is better adapted to their nature; as beef, mutton, &c.,—animal food that is fatted under natural conditions.

There is no one article of food that has caused more disease than pork. Dr. Nichols says that man partakes more or less of the qualities of the animal that he subsists upon; and says, that "in eating pork we swallow the matter which was in process of elaboration, and was destined to become a parts of the brain and nervous system of the hog; consequently we appropriate what would have been the cerebral organ of some swinish nature."

You may be sure that when matter has gone so far towards being converted into the proper essence of the hog, it is not easily turned out of its course. It seems much better to take the pure nutrition furnished us in healthy food, and do for ourselves the whole process of elaboration. One thing is sure—when we feed on pork, we feel stupid, and partake of the sluggish stupidity of the swine.

Pork is not the food for high and pure natures, of noble purposes and earnest thought, of the highest art and profoundest science. Hog is not the aliment of intellectual developments; all men of striking intellectual development have been sparing of animal food, and more especially pork. The hog may be a very respectable animal in his way, but he has no qualities that I am aware of to induce me to eat the creature.

But this is not the worst feature of the case. The process of fatting a hog is a diseasing one; he is shut up in filthy pens and fed upon all the filth and refuse of the farm. A fatted hog seems to be the incarnation of laziness, gluttony, and filth, and the scrofulous disease and utter depravity which these generate. A fatted hog, then, is the epitome of the evils of modern society. I think that Dr. Adam Clark had the right view of the subject when he said, that were he to make a sacrifice to the Devil, he would offer him a fat hog stuffed with tobacco. It is a fact well known to medical men that some hogs are a congeries of scrofulous tumors, and that a considerable portion of their lard is as much corruption as the matter of a tumor. They know all the disgusting and deathly facts of the pathology of the pork market—of ulcerated livers, diseased lungs, and the general corruption of hogs. *Scrofula*—this term is derived from the Latin *scrofa*, a sow; because it is a disease to which swine are especially liable. So when we eat hog we eat scrofa or scrofula. But the hog-advocates will say that they keep the swine for the profit, or as a matter of interest; but it has been ascertained by correct experiment that the corn required to make pork enough to support a man one hundred days, would, if eaten in its pure, original, and far more healthy condition, afford him as much nutriment for four hundred and eighty days, to say nothing of time and trouble lost in feeding the animal. But, says one, the Hog will eat that which no other animal will eat. I will admit that; and that is a good reason why we should not eat him. In fatting a hog, a certain number of bushels of good healthy corn and potatoes are converted into a mass of greasy, and, in many cases, scrofulous pork, with great loss and trouble, while the flesh thus made does not contain one principle necessary to the human constitution which did not exist in a far better form in the vegetable on which it fed. In short, it has been found by an accurate calculation that vegetable food is not merely better, but five hundred per cent. cheaper than the flesh of the swine.

Since the attention of men of science has been turned to organic chemistry, the proportions of nutritive matter in various substances have been accurately ascertained. The following is the result of these inquiries:—Turnips contain 11 per cent. of nutritive matter; beets, 11; carrots, 13; flesh, 25; potatoes, 28; oats, 82; peas, 84; wheat, 85; beans, 86. Corn is about the same as oats and wheat. Thus one hundred pounds of flesh contain but twenty-five pounds of nutritive matter, and seventy-five pounds of water, while the same quantity of potatoes contain twenty-eight pounds of nutritive matter, and wheat eighty-five pounds. But this is not all; the best food is that which contains the materials for muscles, nerves, bones, &c., and the matter for combustion which keeps up vital heat, in proportions. Some may think the author is a Jew, or that his prejudices against pork are unfounded; but it is a duty that I owe to the public as a medical man to raise a warning voice against the effects of hog-eating. Farmers have found of late that it is as much profit to feed their milk and provender to their calves, and keep them until they are five or six months old, as it is to feed it to the swine; and they then have a superior article of animal food altogether.

It would be a great blessing for the community if the Hog would become extinct, and especially in our climate. But for the benefit of those that will raise the Hog, and those that has the sick animal on hand, I shall notice the diseases that his nature, habits, and false conditions subject him to.

STOPPAGE OF THE ISSUES IN SWINE'S LEGS.

Swine are so constituted that they, like the Sheep, do not perspire, consequently the used-up or worn-out particles have to be expelled or carried out of the system through other channels or avenues. For this purpose the Hog is supplied with a number of small expellants, or issues, as they are called, through which a quantity of morbid exhalations pass off, which if retained in the system would cause irritation and disease.

Those issues sometimes become obstructed or stopped from the filthy habits of the swine, and will produce the following

Symptoms—Dizziness, vomiting, inability to move, &c.

Treatment.—Wash the legs in warm soap and water, and open the issues, which may be found on the inside of the forelegs, and keep the hog in a dry, clean place.

MEASLES.

This disease is seldom discovered until the animal is slaughtered, when it is too late to be remedied. It is then that you have a fine mass of scrofula.

Symptoms.—Swine that are troubled with this disease will have a much hoarser voice than usual, their tongues will be pale, and, in bad cases, their skin will be thick-set with blisters, about the size of peas.

Treatment.—Make an infusion of yellow dock, dandelion, and sarsaparilla, and give them the tea to drink, or wet their provender with the same, and keep the animals in a clean, dry, well-ventilated pen, or, what is better, turn them out in a dry pasture.

QUINSY.

This is a disease which Swine are very subject to, and will prevent their feeding, and frequently happens when the animal is half fatted; and this disease will reduce them in a few days to as great poverty in flesh as they were before they were put up to feed.

Symptoms.—A swelling in their throats, which causes great difficulty in swallowing.

Treatment.—Foment the throat over a decoction of bitter herbs, and give ten or fifteen drops of the tincture of belladonna once in fifteen or twenty minutes. The tincture should be dropped on the roots of the tongue, where it can act upon the glands of the throat.

KERNEL.

The distemper called the kernel, is likewise a swelling in the throat. This disease resembles the quinsy, and requires the same treatment, with the addition of a decoction of the roots of common field-narcissus or yellow daffodil.

LOATHING OF MEAT, OR VOMITING UP OF FOOD.

This disease, however, is not mortal, but has the ill effect of reducing swine in their flesh.

Symptoms.—Vomiting, and refusing their food, and losing their flesh.

Treatment.—The food should be light and sparing for a few days, and give

Madder,	2 ounces.
Common salt,	4 "

Mix, and give a tablespoonful once a day.

LETHARGY.

This is a very common disease in the Swine, and is caused by the animals eating too much strong, hearty food, and especially dry food without sufficient water.

Symptoms.—The swine begins to lose flesh without any apparent cause, and is more inclined to sleep than ordinarily. It is common in this distemper for a hog to sleep more than three parts of his time, and consequently he cannot eat, as nature requires him. The hog, in the advanced stages of the disease, will be insensible, and will not move, although you beat him with the utmost violence.

Treatment.—The most certain and approved remedy in this case is,

Wild cucumber root,	4 ounces.
Boiling water,	4 quarts.

Give it to him to drink. This will cause him to vomit. The action on the stomach causes him to become more lively. If this remedy cannot be had, take

Lobelia herb, ... 1 ounce.
Boiling water, . . . 1 quart.

When sufficiently cool, give to the hog to drink, or mix his provender with the same.

If this does not succeed, give the following:—

Bloodroot, ... 2 ounces.
Ginger root, . . . 1 "

Pulverize and mix, and give a tablespoonful each morning in his feed.

PESTILENCE, OR PLAGUE.

This disease prevails epidemically, and is considered very contagious.

Symptoms.—On the attack of this disease, the animal will appear dull, and refuse its food; he soon becomes dizzy and appears unable to move; in some cases he is completely paralyzed; the extremities will become cold, and the muscles rigid.

Treatment.—On the first appearance of this disease, give the following:—

Capsicum, ... 2 ounces.
Ginger, ... 2 "
Boiling water, . . . 2 quarts.

When sufficiently cool, give it in pint doses, once in fifteen or twenty minutes, until reaction takes place, and the extremities become warm. Then give the following:

Mandrake root, ... 1 ounce.
Cream of tartar, . . . 2 "

Pulverize and mix, and divide into three doses, to be given once in six hours until it operates as physic.

INFLAMMATION OF THE LUNGS.

As swine are of a hot nature they are subject to frequent inflammatory attacks, and especially of the lungs.

Symptoms.—This disease may be known by the animal's appearing dull and refusing his food; hot mouth; difficulty of breathing; quick or hurried respiration; cold extremities, and other symptoms of fever.

DISEASES OF DOMESTIC ANIMALS.

Treatment.—On the first appearance of the disease give the following:

Lobelia herb,	4 ounces.
Liquorice,	2 "
Boiling water,	4 quarts.

When sufficiently cool, give one gill at a time, once in fifteen or twenty minutes, until the symptoms abate; then give

Mandrake root,	1 ounce.
Cream of tartar,	2 "

Mix, and divide into six doses, to be given once in six hours, ntil it operates as physic.

Cause.—This disease proceeds generally from want of water. They are never subject to this disease except in the summer time, or when water is wanting. It is frequently to the farmer's expense very greatly, when swine are put up to be fatted, that there is not due care given them in providing water. If they do not have a sufficient quantity, they will surely pine, and lose the benefit of their feed.

BITE OF A VIPER OR MAD DOG.

The symptoms of madness in hogs, which proceed from the bite of a mad dog are, foaming at the mouth and champing with the jaws, starting suddenly at intervals.

Treatment.—Wash the wound immediately with soda-water, as strong as can be made, and give the same to the animal to drink. It has been ascertained of late, that the virus or poison of animals is an acid, and by giving an alkali it will neutralize the acid, and render the poison harmless. Common salt is an antidote for animal poison, when given in large quantities.

SHAKING.

This disease is caused by something that irritates the nerves, and may be known by the shaking or tremor of the whole system.

Treatment.—Take of

Hyssop herb,	4 ounces.
Madder,	2 "
Liquorice,	1 "
Anise-seed,	1 "
Boiling water,	4 quarts.

When cool, give a pint three times a day.

STAGGERS.

This is a very common disease in the hog, and is caused by feeding dry and indigestible food.

Treatment.—When first discovered, take

Lobelia herb,	1 ounce.
Ginger,	2 "
Common salt,	1 "
Boiling water,	4 quarts.

When sufficiently cool, give in half-pint doses until it vomits the hog freely; sometimes it is necessary to repeat the dose several times. Keep the hog on light food for several days.

RECIPES

FOR THE

CURE OF DISEASES OF DOMESTIC ANIMALS.

A CURE FOR WIND-GALLS IN HORSES.

Make a strong decoction of red-oak bark; add to this some strong vinegar, and a little alum in powder. Bathe the parts with this decoction, as warm as possible, twice a day, and bind up comfortably tight with woolen cloths, dipped in the warm decoction.

RECIPE FOR HEAVES.

A correspondent of the "Cultivator" says, after trying various modes of treatment for heaves in horses, he prefers the following:—To feed no hay, but plenty of bright straw, with all the oats that the horse can eat—the oats soaked in cold water—with a pint of oil or flaxseed meal daily in his feed. I have managed horses as above directed, and found it very effectual, and consider this one of the best recipes that I ever tried for heaves in horses.

RECIPE FOR THE THRUSH IN HORSES' FEET.

Simmer over the fire, till it becomes brown, equal parts of honey, vinegar, and verdigris, and apply occasionally, with a feather or brush, to the feet,—the horse, at the same time, on hard dry ground; or if in the stable, all soft dung and straw should be removed.

TO PREVENT SADDLE-GALLS.

Saddle-galls are generally occasioned by an unequal pressure of the saddle, or by the saddle being badly fitted to a horse's back; and, if neglected, they grow into very ugly and troublesome sores.

When these inflamed tumors are first discovered, cold water alone is sufficient to disperse and drive them away, if applied as soon as the saddle is removed; but, when that will not have the desired effect, the back may be washed twice a day in a mixture of

Spirits of turpentine,	2 ounces.
Origanum oil,	1 "
Sweet oil,	4 "

This liniment should be well rubbed in once a day, and bathed in with moderate heat. As in most cases, the evil here is much easier prevented by a little previous caution, than removed after it is produced.

LINIMENT FOR GALLS ON THE HORSE.

One of the most effectual remedies that we have known, for obstinate cases of galled necks and backs, is an application of white-lead paint. If applied in the early stages of the injury, the cure is certain.

TO DRY A COW OF HER MILK.

Circumstances sometimes render it necessary to stop the lactescent action in cows; and, when these occur, all that is absolutely required is, to make a liquid, by pouring into a fresh rennet-bag two quarts of pure well, spring, or rain water; reduce the quantity of the liquor, by boiling briskly, to about one quart, and strain it. Then let it cool to a lukewarm temperature, and give it as a drink to the cow. In forty-eight hours she will be dry. For some days her food should be dry and unsucculent, no water being allowed.

SORE TEATS IN COWS.

P. Hallock, in the "Maine Farmer," gives the following directions for managing of cows that have sore teats:—Take a pail of cold water, and wash and rub the sores well. This cools the teats, and reduces the fever, and the cow will stand perfectly still. After milking, use bitter-sweet ointment, made by simmering the bark of the root in fresh butter.

HOW TO DESTROY LICE ON CATTLE.

Rub them well with rancid lard, whale, or tanner's oil. The "Boston Cultivator," recommends washing the animal a few times with a decoction of red cedar bark. It is said that scattering on them buckwheat flour, or Indian meal, will drive the lice awa

TO KILL LICE ON HORSES OR CATTLE.

Take the water in which potatoes have been boiled, and rub it over the skin. The lice will be dead in two hours, and never will multiply again. A practical farmer says, he has tried various antidotes for lice—among others, the most violent poisons—but has found none so effectual as this.

SALVE TO CURE BURNS, BRUISES, AND SORES.

Olive oil, ... 1 ounce.
White duacula, . . . 2 "
Beeswax, ... 2 "

Let the ingredients be dissolved together, and the salve is formed.

AN EXCELLENT HEALING SALVE.

Rosin,
Beeswax,
Sweet oil.

Melt and mix, stirring until cool. This is a good healing salve for all common sores; but if a more powerful remedy is needed, add to this, when nearly cool, a quantity of red lead, and a little pulverized camphor. This should be spread thin, and renewed once or twice a day.

VOLATILE LINIMENT.

This is a valuable liniment, to be rubbed on the skin as an external stimulant, and is good for sprains, rheumatisms, spasms, and lameness in general. After rubbing it well in, which should be continued from twenty minutes to half an hour, flannel should be wrapped around the affected parts. Volatile liniment is made by mixing equal quantities of spirits of hartshorn and sweet oil; by adding to this mixture a teaspoon or two of laudanum, the preparation will be much improved in its efficacy in relieving pain.

BLOODY URINE IN CATTLE.

Powdered alum,	. . . .	1 ounce.
" rosin,	. . .	2 "
" salt,	. . .	1 "
" loaf sugar,	. .	4 "

Mix, and give three spoonfuls each morning, mixed with meal, and the cure is sure.

COLIC IN CATTLE.

Brandy,	. . . .	1 pint.
Laudanum,	. . .	½ ounce.
Water gruel,	. . . .	2 quarts.

Divide into two doses; give one half, and if relief is not obtained in fifteen minutes, give the remainder.

GARGET, OR SWELLED BAG.

Hogs' lard,	. . . .	4 ounces.
Henbane, when in bloom,	.	3 "

Cut fine, and simmer together over a slow fire until crisp; strain out with a smart pressure, and apply the ointment.

ANOTHER CURE FOR GARGET.

Take raw linseed oil and anoint the cow's bag as soon as the swelling appears. Two or three applications are a sure cure.

CURE FOR THE HORN DISTEMPER.

Salt,	. . . . .	4 ounces.
Black pepper,	. . .	2 "
Ginger,	. . . .	4 "
Soot,	. . . .	6 "
Tar,	. . . . .	8 "

Mix, and make twelve balls, each the size of a hen's egg; three balls to be given at a time, at intervals of three days, until all are given, and the cure is sure, if given in time. I have known this remedy these twenty years, and seldom seen it fail.

CURE FOR HOOF-AIL.

Sulphate of soda,	8 ounces.
Ginger,	2 "
Molasses,	1 pint.

Dissolve in three pints of warm water, and give at once. This will physic the animal. Then apply a few drops of nitric acid, diluted in twenty times its weight of water, to the part of the hoof affected; at the same time keep the foot free from dirt. If very sore, apply a bran poultice. The acid should be applied every two or three days. This is found to be an effectual remedy. Another application for the foot is

Blue vitriol,	1 ounce.
Soft grease,	1 "
Salt,	1 "

Mix well together. After trimming the parts of the hoof affected, take a small brush, or a stick with a piece of rag rolled round the point, and apply the remedy to the diseased parts. On cattle use it once a day for four or five days. Sheep may be cured by two or three applications, in a few days.

HOOVE IN CATTLE.

Nux vomica, powdered,	1 teaspoonful.
Rain water,	1 pint.

To be given at a dose. This is a safe and speedy cure.

SORE TEATS IN COWS.

Take bitter-sweet bark, from the root, wash and simmer it with a quantity of lard, until it is very yellow; and, when cool, strain and apply it to the parts that are swollen, two or three times a day, until the udder and teats are perfectly soft and free from kernels. This has been used with great success.

CURE FOR STIFLE.

Powdered alum,	2 ounces.
" salt,	1 "
The whites of	3 eggs.

Beat the whites of the eggs to a stiff froth, then add the alum. Coat over the stifle joint, and hold near enough to dry in. I have never known two or three applications to fail of entire cure, in the worst cases, but generally one is sufficient.

KICK IN THE STIFLE.

If the joint-water runs, apply the balsam of copaiba. This is a safe and effectual remedy.

SCOURS IN CATTLE.

Give a pint of good, strong rennet to a cow three times a day; or three tablespoonfuls to a sheep, three times a day. This remedy seldom fails.

SORE NECKS IN OXEN.

Use ointment made of lard and beeswax; or make a strong wash of white-oak bark, and apply it twice a day.

WOUNDS IN CATTLE.

Sulphate of zinc,	1 ounce.
Common salt,	1 "
Rain water,	1 pint.

This is a cure for common sores, and especially where there is much inflammation.

WARTS ON COWS' TEATS.

These may be cured by simply washing them in alum-water. A strong decoction of black-oak bark, applied twice a day, after milking, for two or three weeks, is also an effectual remedy.

COLIC IN HORSES.

Dissolve in a quart of pure water as much salt as will thoroughly impregnate the liquid, and drench the animal thoroughly until you discover symptoms of relief.

Another Remedy for Colic.

Drench your horse with sage tea,—it should be very strong. To every quart of the tea add one ounce of paregoric.

COUGH IN HORSES.

The boughs of cedar or white pine, cut fine and mixed with the grain, and given to the horse, are a good remedy.

CRAMP IN HORSES.

Apply hot fomentations to the limbs, and give a dose of cathartic medicine.

CORNS ON HORSES' FEET.

Dig the corn out, and apply hot tar. If the horse is flat-footed be careful about paring the heel—only rasp it carefully, and as much as possible avoid having the shoes press on the heels.

FOUNDER IN HORSES.

Give from one to two quarts of strong brine; be careful not to let him drink too much; then anoint round the edges of his hoofs with spirits of turpentine, and the cure will be sure and speedy.

Another Cure for Founder.

Take a tablespoonful of pulverized alum, pull the horse's tongue out of his mouth as far as possible, and throw the alum down his throat; let go of his tongue, and hold up his head until he swallows. In six hours' time, if applied immediately after the accident, the horse will be fit for moderate service.

Another Cure for Founder.

Mix one pint of the seed of the common sunflower in the animal's food, as soon as you discover symptoms of founder, and it will give immediate relief. This is a good remedy.

FISTULA IN HORSES.

Alcohol,	1 pint.
Turpentine,	½ "
Indigo,	1 ounce.

Mix, and apply a small quantity every other day.

CURE FOR GLANDERS.

In the first place, put a rowel or seton of poke-root between the jaws and in the breast; then procure one gallon of fresh tar; next fix a small swab on a stick long enough to insert it as high up as the eyes, then insert the tar into the nostrils in this way twice a day until you make a complete cure.

CURE FOR HEAVES IN HORSES.

Feed no hay, and add two parts of Indian meal to one of shorts, adding to each feed a tablespoonful of ginger. Horses managed in this way will perform all moderate work as well as if sound.

Another Cure for the Heaves.

Tincture of aromatic sulphur acid,	1 drachm.
Water,	1 pint.

To be given as a dose. Most horses will drink it from a bucket. In the meantime the horse should be put on a course of alterative medicine; such as

Powdered ginger,	2 ounces.
" gentian,	2 "
" salt,	3 "
" cream of tartar,	2 "
" liquorice,	2 "
" elecampane,	1 "
" caraway seed,	2 "
" balm of gilead buds,	2 "

Mix, and give one ounce every night and morning in the food. All healthy conditions should be observed and fulfilled. As soon as the horse is considerably improved, the aromatic tincture should be omitted; and instead of giving an ounce of the alterative as a dose, give half-ounce doses.

Another Cure for Heaves.

Sumach bobs,	3 pounds.
Ginger,	1 "
Mustard seed,	1 "
Rosin,	1 "
Air-slacked lime,	1 "
Cream of tartar,	6 ounces.

Mix, and divide into thirty parcels, giving one every morning in a cut feed. Feed good bright straw, and keep all dust and hay away. The best fodder for heavey horses is good, bright cornstalks.

HOOF-AIL IN HORSES.

Apply blue vitriol, and put on a tarred rag to keep out the dirt.

CURE FOR THE FOOT-EVIL IN HORSES.

Wash the horse's foot well with warm soap-suds; wipe it dry with a cloth; then take

Common salt,	2 ounces.
Copperas, pulverized,	1 "
Soft soap,	½ pint.

Mix them well; spread upon a thick cloth, and apply it to the foot. Then confine it with a cloth; let it remain twelve hours; then take it off; wash as before, and it is a sure cure. The disease will not spread any afterwards. If, when mixed, the compound is too stiff, moisten with water.

HIDEBOUND HORSES.

The cure of hidebound horses consists in restoring to healthy action whatever organs are diseased. The general health must be improved ere the coat will assume its natural sleek appearance, and soft, pliant feel. The secretory vessels that nourish and support the hair must be free and active. If no particular disease can be detected about the animal, give plenty of nutritious food and moderate exercise; if this does not improve the condition of the animal, give the following alterative in his food night and morning:—

Powdered sassafras bark,	3 ounces.
Salt,	3 "
Powdered bloodroot,	2 "
" golden seal,	3 "
" ginger root,	2 "
Oatmeal,	1 pound.

Mix, and divide into twelve parts.

LAMPASS IN HORSES.

The first thought of our farmers is to take the animal to be burned. This arises from ignorance and superstition.

This disease is caused by the bars becoming irritated and inflamed, and may be cured by the simple application of

Loaf sugar,	4 ounces.
Alum,	2 "
Salt,	1 "

Pulverize and mix, and rub the gums of the animal two or three times a day, and the cure is mild, safe, and effectual.

TO PURGE A HORSE.

White chalk, powdered,	1 ounce.
Vinegar,	1 pint.

Shake well, and give to the horse when effervescing. It generally operates in fifteen minutes, and affords almost instant relief.

CURE FOR RINGBONE.

Alcohol,	1 pint.
Origanum oil,	1 ounce.
Gum camphor,	1 "

Mix, and apply the medicine once a day; bathe it in moderately by applying mild heat.

ANOTHER RINGBONE CURE.

Spirits of turpentine,	1 pint.
Lamp oil,	4 ounces.
Oil spike,	1 "

Mix, and apply to the parts affected night and morning, rubbing it well into the hair around the edge of the hoof.

CURE FOR A SPAVIN.

Alcohol, ... 1 pint.
Gum camphor, . . . 2 ounces.

This medicine should be applied till the hair starts, but not to blister severely; then let the horse rest a few days, and repeat the operation.

ANOTHER SPAVIN CURE.

Oil vitriol, . . . 1 ounce.
" origanum, . . . 1 "
" cedar, ... 1 "
Oil of olives, . . . 1 "
Spanish flies, . . . 1 "
Spirits turpentine, . . 3 "

CURE FOR A SPLINT.

Alcohol, ... 1 pint.
Oil of cedar, . . . 1 ounce.

To be well rubbed in, and bathed well with a hot brick.

FOR SWEENY, OR SHOULDER-SPRAIN,

Spirits of turpentine, . . 3 ounces.
Balsam of sulphur, . . 2 "

Mix, and apply every other day, and bathe in well with moderate heat.

SCRATCHES IN HORSES.

Wash the parts affected with castile soap; then make an ointment of

Gunpowder, . . . 1 ounce.
Spirits of turpentine, . . 1 "
Pig's-foot oil, . . . 4 "

Mix, and rub it in thoroughly with the hand once a day.

If the heels are much swollen and hot, apply a poultice of the following :—

Slippery elm, powdered, . . 8 ounces.
Salt, ... 2 "
Hot water, . . . 1 quart.

When cool, apply it in the form of a poultice; continue this until the inflammation subsides.

STIFLE, OR SHOULDER-SPRAIN.

Take white-oak bark boiled strong; to every quart of this decoction add one tablespoonful of salt. Bathe, while warm, several times a day. A cure is commonly effected in two or three days.

Another.

Balsam of fir, ... 4 ounces.
Oil of spike, ... 3 "
Sweet oil, ... 2 "

Mix, and apply to the parts affected, and bathe it in well with a hot brick. To be repeated until a cure is affected.

CURE FOR WIND-GALLS.

Glycerine, ... 2 ounces.
Tannin, ... 2 drachms.
Water, ... 4 ounces.

Mix, and apply two or three times a day.

WORMS IN HORSES.

The following is very effectual, and I consider it one of the best vermifuges now in use.

Castor oil, ... 12 ounces.
Wormseed oil, ... 1 "
Oil of tansy, ... 3 drachms.

To be given on an empty stomach, followed by bran-mashes and salt.

YELLOW WATER.

This is generally the consequence of hard fare and a scanty allowance of food. In this case, give plenty of good sweet oats; and if this does not improve the condition of the animal, give

Quinine, ... 2 drachms.
Golden seal, ... 2 ounces.
Ginger, ... 2 "

Mix, and give a tablespoonful once a day.

COSTIVENESS IN SHEEP.

This disease should be removed by change of diet, if possible; if not, give castor oil, a tablespoonful once in twelve hours, until it has the desired effect. This may be assisted by an injection of warm, weak suds and molasses.

COLIC IN SHEEP.

This disease may be cured by giving
Castor oil, ... 4 ounces.
Ginger, ... 1 drachm.
Paregoric, ... 1 ounce.

DIARRHEA IN SHEEP.

Prepared chalk, ... 1 ounce.
Catechu, ... $\frac{1}{2}$ "
Ginger, ... 2 drachms.
Opium, ... $\frac{1}{4}$ "

Pulverize and mix them with half a pint of peppermint water; give two or three tablespoonfuls morning and night to a grown sheep, and half the quantity to a lamb.

DYSENTERY IN SHEEP.

Give plenty of mucilaginous drinks, and give
Peppermint herb, . . . 1 ounce.
Rhubarb, ... $\frac{1}{2}$ "
Soda, ... $\frac{1}{2}$ "
Steep in boiling water, . . . 1 quart.

strain and add:—
Loaf sugar, ... $\frac{1}{2}$ pound.
Laudanum, ... 1 ounce.

Give a tablespoonful three times a day.

FOULS IN SHEEP.

Sheep are much less subject to this disease than cattle; but are liable to it if kept in dirty, wet yards, or on moist, boggy grounds. It is an irritation of the integuments of the clefts of the foot, slightly resembling incipient hoof-ail, and occasions lameness; but it produces no serious constructional disorganization, and sometimes disappears without treatment. It is not contagious, and appears in the wet weather of spring and fall, instead of the dry, hot period of summer when the hoof-ail rages most. A solution of blue vitriol, or a little spirits of turpentine, followed by a coating of tar, soon cures the disease.

FOOT-ROT IN SHEEP.

Take one half pound gunpowder; one half pound burnt alum, finely powdered; to which add one ounce of sulphuric acid, and three gills soft water. Mix the whole thoroughly together in a glass or earthen vessel, and apply immediately to all the feet that are diseased; clean the feet well in soap and water before the application of medicine. This is a sure cure.

SCAB IN SHEEP.

The Scab is a cutaneous disease, caused by minute insects, called *acari*. This may be cured by the following:

Black antimony,	1 ounce.
Nitre in powder,	2 "
Sulphur,	4 "
Ginger,	1 "
Lard,	1 pound.

Pulverize and mix; anoint the sheep thoroughly.

SORE MOUTH IN SHEEP.

This is supposed by some to be caused by sheep eating, in the winter season, noxious weeds; for it is at this period of the year generally that they are most subject to it.

Remedy.—Tar applied freely to the mouth.

STRETCHES IN SHEEP.

The following is the remedy :—

New milk,	$\frac{1}{2}$ pint.
Soap,	$\frac{1}{2}$ "
Molasses,	$\frac{1}{8}$ "

Mix, and divide into two doses; give one, and wait fifteen or twenty minutes, and if relief is not obtained, give the remainder.

STAGGERS IN SHEEP.

Pour spirits of turpentine into the ears of the sheep, and give catarrh snuff. This is a good remedy, and seldom fails.

TICKS IN SHEEP.

The smoke is a sure remedy for ticks. The sheep should be enveloped in a sack all but the nose, and that sack communicate with the fumes of tobacco.

WORMS IN SHEEP.

It is a well-known fact that sheep are sometimes troubled with worms in the head, to the great annoyance of the whole flock. The most effectual remedy that I have ever known is, to take honey, dilute with a little warm water, and inject into the nose freely with a syringe.

BLACK TEETH IN SWINE.

This disease is caused by keeping the swine from the ground and under unnatural conditions.

The remedy is, to turn the hogs out, and let them have free access to the ground and pure air.

COSTIVENESS IN SWINE.

This may generally be remedied by plenty of vegetables or wheat bran. If this fails, a little sulphur may be given.

KIDNEY-WORM IN SWINE.

This disease may be generally known by the animal appearing weak across the loins, and sometimes by a weakness in both hind legs. Rub the loins with turpentine, and soak the animal's corn in lye or wood ashes, or strong soap-suds.

LICE ON HOGS.

If hogs be well fed and kept clean, they will not get lousy. Their bedding should be renewed at least once a week. Filth and insufficient food are the parents of lice.

Remedy.—A moderate portion of sulphur given in the food will soon destroy them. A good coating or two of soft-soap is effectual.

CHOLERA IN SWINE.

Flaxseed,	2 quarts.
Water,	10 gallons.

Boil till the seed is thoroughly cooked; let it stand till cool, then give it to the hogs as fast as they will drink it; turn it down if they refuse to drink; repeat the dose for a week or so, and it will effect a cure.

STAGGERS IN SWINE.

Pour soft oil on the issues of the legs, and rub them well; then give as much rum and pepper as you can get them to take with a spoon. This has cured cases where they were in the last stages of the disease.

THUMPS IN HOGS.

At the South a disease prevails in hogs, called the Thumps, which is very fatal. In the southern portion of the State some have lost over a hundred, and they are still dying.

The remedy is, give to each hog one tablespoonful of spirits of turpentine every other day, until a cure is effected. One or two doses generally are sufficient.

TAIL DISEASE IN HOGS.

The tails of young pigs frequently drop or rot off, which is attended with no further disadvantage to the animal than the loss of the member. The remedies are, give bloodroot and salt in the food of the dam before the pig is weaned, or to the pigs afterwards.

CANKER IN THE THROATS OF FOWLS.

This disease prevails in some sections of the country to an alarming extent. It appears to be a disease of the windpipe, which sometimes fills up and causes the fowls to die. The mouth is full of cankered spots.

The remedy is this: Black pepper, salt, and vinegar. Swab the throat well, and the cure is sure.

CHOLERA IN FOWLS.

This is a very prevalent disease in California, and has carried off thousands during the spring and summer.

Preventive.—Put two ounces of oxyd of iron into one pint of quick-lime, and this into two gallons of water; to be kept cool; let the fowls drink freely of this as a common water. This is said to be a sure preventive.

Treatment.—Give two or three drops of tincture of ginger once in fifteen minutes, until a cure is effected.

GAPS IN CHICKENS.

Spirits of turpentine will be found a sure cure for gaps in chickens.

Put one drop of spirits of turpentine into half a teaspoonful of water, sweeten with sugar, and give to each chicken.

ROUP IN FOWLS.

This disease is highly contagious, and is the most troublesome disease that we have to contend with in the management of poultry.

Treatment.—Take equal parts of charcoal and hyssop, mix with butter, and give a piece of the size of a hazelnut, night and morning; wash the heads with cold water, and keep their nostrils clean.

VERMIN IN FOWLS.

Oil the roosting-poles, once in six or eight weeks, with fish-oil, and dust the hens with fine sulphur, and the lice will soon disappear.

HOOF OINTMENT.

Balsam of fir,	4 ounces.
Spirits of turpentine,	1 "
Rosin,	2 "
Tallow,	4 "

Melt slowly over a slow fire, and, when thoroughly melted, remove from the fire, and stir until cool.

CONDITION POWDER.

Antimony,	1 ounce.
Cream of tartar,	4 "
Sulphur,	4 "
Liquorice,	2 "
Fenugreck,	$\frac{1}{2}$ "

Mix, and give a tablespoonful once a day. This is the Condition Powder that is sold so extensively all over the country; and all of the powders sold amount to about the same thing.

CONDITION POWDER.

Powdered blood root,	3 ounces.
" skunk cabbage,	2 "
" ipecac,	1 "
" liquorice,	2 "
" fenugreck,	1 "

Mix, and give a tablespoonful three times a day. This is one of the most valuable Condition Powders in use, and is free from all minerals.

THE CELEBRATED SPAVIN AND RINGBONE OINTMENT.

(*Prepared and sold at a high price in Otsego County, N.Y.*)

Lard,	1½ lb.
Tincture of Spanish flies,	1 ounce.
Gum euphorbium,	2 "
Tincture of iodine,	2 "
Spirits of turpentine,	2 "
Quicksilver,	½ "
Corrosive sublimate,	2 drachms.

Mix the quicksilver and sublimate together, and then the rest of the articles.

POISONS AND THEIR ANTIDOTES.

There is nothing that could be inserted in this work that may be of more use to the farmer than a list of poisons and their antidotes, seeing that cases of poisoning are so common and so mortal in their effects. The knowledge that I am about to impart may save the life of some of his family, or that of a valuable animal; as the antidotes are the same for poisons whether in the stomach of man or animal; but the medicines or antidotes should be given in much larger doses in animals than in man.

When any poison has been swallowed by man, the first thing to be done is to drink freely of milk or mucilaginous drinks, and then tickle the fauces with a feather or the finger, by running it down the throat. This will excite vomiting without the loss of time to wait the action of an emetic.

When a patient has taken any mineral poison, he should drink freely of the whites of eggs beat up in water, or wheatflour starch, sweet milk, cream, or oil.

As vomiting cannot be produced in the Horse by any known medicine, the drug or poison must be neutralized in the horse's stomach, or changed, chemically, by forming new combinations that will be harmless, or possess such properties that they are not dangerous, and can be safely removed from the system.

LIST OF POISONS AND ANTIDOTES.

When the nature of the poison is unknown, give magnesia, pulverized charcoal, or excite vomiting as above directed.

[Poisons in *Italic*; antidotes in Roman.]

Oxalic acid. Magnesia, lime.
Nitric acid (Aquafortis). Lime, chalk, magnesia, soda.
Muriatic acid. Soda, lime, magnesia, potassa, ammonia.
Prussic acid. Ammonia, carbonate of potassa in solution, water, stimulants.
Citric acid. Ammonia, magnesia, chlorine, carbonate of potassa in solution, soda and lime.
Arsenic or Arsenious acid. Calcined magnesia, hydrated peroxyd of iron.
Oil of Vitriol. Milk, oil soda, magnesia, lime, chalk.
Tartaric acid. Lime, saleratus, soda, plaster from the ceiling.
Verdigris. The whites of eggs, cream or milk.
Sugar of Lead. Sulphate of soda, sulphate of magnesia, potassa in solution.
Morphine. Strong tea, coffee, stimulants, dash of cold water.
Aconite. Tannic acid, green tea, iodine.
Sulphate of Zinc. Common table salt, charcoal, chlorine.
Sulphuric Ether. Ammonia by inhalation.
Hartshorn. Vinegar, lemon juice.
Bitter Almonds. Vinegar, lemon juice, ammonia, charcoal.
Meadow Pimpernel. Charcoal, green tea.
Antimony. Astringents, tannic acid, alkalies.
Emetic, Tartar. Astringents, yellow-bark, green tea.
Wine of Antimony. Tannic acid, infusion of green tea.
Dogsbane. Charcoal, soda.
Silver. Common table salt.
Nitrate of Silver. Common table salt.
Elixir Vitriol. Magnesia, lime, chalk, soda.
Arsenic. Magnesia, hydrate of iron.
Wake Robin (Wild Turnip). Charcoal.
Belladonna. Chlorine, iodine, stimulants, lime water, vinegar.
Chloride of Gold. Sulphate of iron, mucilage.
Bee-sting. Solution of ammonia, solution of common salt, tincture of iodine.
Bichromate of Potash. Carbonate of soda, saleratus, carbonate of potassa.

Subnitrate of Bismuth. Mucilage, milk, eggs.
The bite of a Serpent. Alcohol, ammonia, asclepias verticulata.
Bromate of Potassa. Whites of eggs, starch.
Bryony. Bromine, iodine, chlorine.
Calomel. Gluten, gold, iodine.
Quick-lime. Mineral soda-water, effervescing draughts.
Camphor. Common salt, mustard seed, emetics.
Spanish Flies. Whisky, ammonia, soda.
Celandine. Emetics.
Wormseed. Emetics.
Chloroform. Ammonia by inhalation, galvanic shock, a strong solution of salt and water, bathe spine with salt solution.
Cicuta (poison Hemlock). Emetics.
Cocculus Indicus (Indian cockle). Bromine, chlorine, iodine
Colocynthine. Bromine, chlorine, iodine.
Jalapa (Jalap). Bromine, chlorine, iodine.
Corrosive Sublimate. Albumen, the whites of eggs, gluten, gold-dust, iron-filings.
Gnat-bite. Solution of ammonia.
Creosote. Whites of eggs; milk, flour and water.
Cyanide of Potassa. Sulphate of iron in solution.
Foxglove (Digitalis). Infusion of yellow-bark, stimulants, tannic acid, green tea.
Swamp Leatherwood. Chlorine, bromine, iodine.
Squirting Cucumber. Bromine, chlorine, iodine.
Spurred Rye (Rye smut). Charcoal.
Salts of Iron. Carbonate of soda, magnesia, mucilage.
Muriate Tincture of Iron. Subcarbonate of soda, saleratus.
Salts of Gold. Sulphate of iron, mucilage.
Yellow Jessamine. Ammonia, charcoal.
Hedge Hyssop. Charcoal.
Black Hellebore (Itchweed). Charcoal.
Marsh Pennywort. Charcoal.
White Henbane. Charcoal, vinegar, ammonia.
Black Henbane. Vinegar, ammonia, iodine.
Hyoscyamus. Vinegar, ammonia, iodine.
Mercury. Whites of eggs, gluten, iodine.
Iodine. Gluten, wheat flour starch.
Iodide of Potassium. Gluten, wheat flour starch.
Ipecacuanha. Bromine, chlorine, iodine.
Iron, and its salts. Carbonate of soda, carbonate of magnesia.

Laurel. Inhalation of ammonia, chlorine, chloroform.
Lead, and its salts. Diluted sulphuric acid, iodide of potassa, sulphate of soda, sulphate of magnesia.
Morphia, and its salts. Astringents, charcoal, coffee, green tea.
Blue Parrot Fish. Charcoal.
Mushrooms. Charcoal.
Conger Eel. Charcoal.
Muscle Fish. Charcoal.
Daffodil. Charcoal.
Gad Fly. Solution of Ammonia.
Oil of Tobacco. Charcoal.
Oil of Turpentine. Ammonia.
Opium, and its preparations. Infusion of galls, astringents, coffee, magnesia, chlorine, charcoal, iodine, green tea, vinegar.
Poison Parsnip. Charcoal.
Copper. Charcoal, ammonia.
Phosphorus. Magnesia, mucilage, ammonia, cold water.
Poke. Charcoal.
Cubebs. Charcoal.
Iodine. Sulphuric acid (diluted), sulphate of soda, sulphate of magnesia, albumen, milk.
Caustic Potash. Oils, vinegar, lemon juice.
Bicarbonate of Potash (Saleratus). Lemon juice, vinegar.
Poppy. Infusion of galls, tannic acid, charcoal, ammonia, green tea.
Wild Orange. Ammonia, inhaled; chlorine, inhaled; chloroform.
Cherry Laurel. Dash of cold water; ammonia, inhaled; chlorine, inhaled; chloroform, inhaled.
Black Cherry. Dash of cold water; ammonia, inhaled; chlorine inhaled; chloroform, inhaled.
Cluster Cherry. Ammonia, inhaled; chlorine, inhaled; chloroform, inhaled.
Wild Cherry. Dash of cold water; ammonia, inhaled; chlorine, inhaled; chloroform, inhaled.
Prussic Acid. Dash of cold water; ammonia, inhaled; chlorine, inhaled; chloroform, inhaled.
Poison Vine. Charcoal.
Putrid Animal Matter. Ammonia, tonics, scutellaria lateriflora (skull-cap).

Poison Oak, or Sumach. Charcoal.
Oleander. Charcoal.
Locust Tree. Charcoal. (I have known many a horse poisoned by eating the bark of the common locust tree.)
Rue. Charcoal.
Mad Dog Saliva. Nitrate of silver, ammonia, scutellaria lateriflora (skull-cap).
Spanish Mackerel. Charcoal.
King Fish. Charcoal.
Bonetta. Charcoal.
Carbonate of Soda. Vinegar.
Bitter-sweet. Charcoal.
Mountain Ash. Charcoal.
Pinkroot. Charcoal.
Gamboge. Charcoal.
Chloride of Tin. Whites of eggs, milk, flour.
St. Ignatius' Bean. Bromine, chlorine, iodine, prussic acid, potassa, chloroform.
Dog Button (Nux Vomica). Bromine, chlorine, iodine, prussic acid, potassa in solution, chloroform.
Strychnia. Bromine, chlorine, iodine, prussic acid, potash, saleratus or soda in solution, chloroform.
Sulphate of Indigo. Magnesia, milk, oil.
Skunk Cabbage. Charcoal.
Tansy Oil. Charcoal.
Muriate of Tin. Whites of eggs, milk, flour starch.
Upas Tree. Charcoal, ammonia.

EXPLANATIONS OF MEDICAL TERMS.

Absorbent. In anatomy, a vessel which imbibes; in medicine, any substance which absorbs or takes up the fluids of the stomach and bowels.
Abscess. A cavity containing pus, or a collection of matter, as a common boil or felon, or any swelling that has come to a head.
Abortion. A miscarriage, or producing a child before the natural time of birth.
Ablution. Cleansing by water; washing of the body externally.
Abdomen. The lower part of the belly.
Abrasion. The act of wearing or rubbing off.
Accoucher. One who assists a female in child-birth.
Accuminate. Taper-pointed.
Acetabulum. The socket that receives the head of the os femoris, or thigh-bone.
Acid. Sour, sharp, or biting.
Acrid. Sharp; pungent; bitter; biting to the taste.
Actual Cautery. Burning or searing with a hot iron.
Acupuncture. Pricking the parts affected with a needle.
Acute. Sharp; ending in a point. Acute diseases are of short duration, attended with violent symptoms—opposite to chronic.
Adhesive. Sticky; tenacious; tending to adhere.
Adhesive Plaster. Sticking plaster.
Adhesive Inflammation. That kind of inflammation which causes adhesion.

Adjuvant. An assistant; a substance added to a prescription to aid the operation of the principal ingredient or basis.
Adult Age. Grown to full size; manhood or womanhood.
Affection. Disorder; disease; malady.
Affusion. The act of pouring upon or sprinkling with a liquid substance.
Albumen. The white of eggs; a principle of both animal and vegetable matter.
Alkali. A substance which is capable of neutralizing acids, and destroying their acidity. Potash, soda, &c., are alkalies.
Alimentary. Having nourishing qualities as food.
Alimentary Canal. The intestines by which aliments are conveyed through the body, and the useless parts evacuated.
Alterative. A medicine which gradually changes the condition of the system, restoring healthy functions without producing sensible increase of the evacuations.
Alveoli. The sockets in the jaw in which the teeth are fixed.
Alvine. Pertaining to the intestines.
Amaurosis. A loss or decay of sight without any visible defect in the eye, except an immovable pupil.
Amenorrhea. An obstruction of the menstrual discharge.
Ament. Having flowers on chaffy scales, and arranged on slender stalks.
Anasarca. Dropsy of the skin and flesh.
Antiscorbutic. A remedy for scurvy.
Antiseptic. That which resists or removes putrefaction or mortification.
Antispasmodic. That which relieves spasm, cramps, and convulsions.
Antisyphilic. Remedy against syphilic or venereal disease.
Anchylosis. Stiffness of a joint.
Aneurism. A soft tumor, arising from the rupture of the coats of an artery.
Angina Pectoris. A peculiar nervous affection of the chest.
Angina Tonsillaris. Inflammation of the tonsils.
Angina Trachealis. Inflammation of the wind-pipe.
Annual. Yearly. An annual plant grows from the seed to perfection and dies in one season.
Annulated. Surrounded by rings.
Anodyne. A medicine which allays pain and produces sleep.

Anti-acid. A substance to counteract acids, as an alkali.
Anthelmintic. A worm-destroyer; worm medicine.
Antibilious. Counteracting a bilious complaint.
Antidote. A protective against or remedy for poison, or any thing noxious to the stomach.
Antidysenteric. A remedy for dysentery.
Anti-emetic. A remedy to check or allay vomiting.
Antilithic. A medicine to prevent or remove urinary calculi or gravel.
Antimorbific. Anything to prevent or remove disease
Aperient. A gentle purgative; a laxative.
Apex. The top or summit; the termination of any part of a plant.
Aroma. The fragrance of a plant or other substance; any agreeable smell.
Aromatic. A fragrant, spicy plant, drug, or medicine.
Arthrodia. A joint movable in every direction.
Ascarides. Pin worms, or thread worms, always found in the lower part of the bowels, or anus.
Ascites. Dropsy of the belly.
Assimilation. The conversion of food into the fluid or solid substance of the body.
Asthmatic. Troubled with asthma or difficulty of breathing.
Astringent. Binding; contracting; strengthening; opposed to laxative.
Atony. Debility; want of tone; defect of muscular power.
Atrophy. A wasting of flesh and loss of strength, without any sensible cause.
Axillary. Pertaining to the arm-pit.
Axillary Gland. A gland situated in the arm-pit.
Barnacles. An instrument to put on the nose of a horse to confine him.
Balsamic. Medicines employed for healing purposes.
Belching. Ejecting wind from the stomach.
Biennial. In botany, continuing for two years, and then perishing; as plants whose roots and leaves are formed the first year, and which produce fruit the second.
Bifurcation. Division into two branches.
Bronchial. Belonging to the ramifications of the windpipe in the lungs.
Bolus. A soft mass of anything medicinal.
Box. Stall, or place of confinement.

Bulbous. Round, or roundish.
Cachexia. A bad condition of the body; where the fluids and solids are vitiated, without fever or nervous disease.
Cadaverous. Having the appearance or color of a dead human body; wan; ghastly; pale.
Calculi. The gravel and stones formed in the kidneys or bladder, or in any part of the body.
Callous. Hard, or hardened, as an ulcer.
Callus. Bony matter which forms about fractures.
Caloric. The element of heat.
Capillary. Resembling a hair; a minute vessel.
Capsule. The seed-vessel of a plant.
Carbon. Charcoal.
Carbonic Acid. A combination of two parts of oxygen with one part of carbon.
Carminative. A medicine which allays pain, and expels wind from the stomach and bowels.
Cartilage. Gristle; a substance similar to, but softer than bone.
Catamenia. The monthly evacuations of females; the menses.
Cataplasm. Poultice.
Cathartic. A purgative; a medicine that cleanses the bowels.
Catheter. A tubular instrument for drawing the urine.
Caustic. Any substance which burns or corrodes the parts of living animals to which it is applied.
Cautery. A burning, searing, or corroding any part of an animal body.
Cellular. Consisting of or containing cells.
Cerebellum. The hinder and lower part of the brain.
Cespitose. Growing in tufts.
Cespitous. Pertaining to turf; turfy.
Choleric. Easily irritated.
Chronic. Continuing a long time; inveterate; the opposite of acute.
Chyme. The modification which food assumes after it has undergone the action of the stomach.
Cicatrix. A scar remaining after a wound.
Clyster. An injection; a liquid substance thrown into the lower intestines.
Coagulation. Changing from a fluid to a fixed state.
Coalesce. To grow together; to unite.

Colliquative. Weakening, as sweating; applied to excessive evacuations, which reduce the strength of the body.
Coma, or *Comatose.* Lethargic; strongly disposed to sleep.
Combustion. Burning with a flame.
Concave. Hollow. A concave leaf is one whose edge stands above the disc.
Concrete. A compound; a united mass.
Confluent. Flowing together; meeting in their course.
Congenital. Begotten or born together.
Conglobate. Formed into a ball.
Connate. United in origin; united into one body.
Constipation. Obstruction and hardness of the contents of the intestinal canal.
Constriction. A contraction or drawing together.
Contagious. Catching; or that which may be communicated.
Contusion. A bruise.
Convalescent. Recovering health and strength after sickness or debility.
Convoluted. Rolled together, or one part on another.
Cordate. Having the form or shape of a heart.
Cordial. A medicine which increases the strength and raises the spirits when depressed.
Coriaceous. Tough, or stiff; like leather.
Corolla. The inner covering of a flower.
Corroborant. A medicine that strengthens the human body when weak.
Corrosive. That which has the quality of eating or wearing away gradually, as an acrid poison of great violence.
Cortex. The bark of a tree or plant.
Cranium. The skull.
Crassamentum. The thick red part of the blood.
Crepitus. A sharp, abrupt sound.
Cutaneous. Belonging to the skin.
Cuticle. The scarf-skin, or outer skin.
Decarbonize. To deprive of carbon.
Decoction. The virtues of a plant extracted by boiling.
Delirium. Confusion of the intellect; wildness or wandering of the mind.
Demulcent. A mucilaginous medicine that sheathes the tender and raw surfaces of diseased parts.
Depletion. Blood-letting.

Deobstruent. Any medicine which removes obstructions and opens the natural pores of the body.
Depuration. The cleansing of impure matter.
Derm. The natural covering of an animal; the skin.
Detergent. A medicine that cleanses the vessels of the skin from offending matter.
Diagnosis. The distinction of one disease from another by its symptoms.
Diaphoresis. Increased perspiration or sweat.
Diaphoretic. A medicine which produces sweating.
Diaphragm. The midriff, or muscular division between the chest and belly.
Diarrhea. A morbidly frequent evacuation of the intestines.
Diathesis. The disposition of the body, good or bad.
Dichotomous. Regularly divided by pairs from top to bottom.
Digest. To disolve in the stomach; or, in medicine, to make a tincture.
Digitate. Divided like fingers.
Diluent. That which thins, weakens, or reduces the strength of liquids.
Diluting. Weakening.
Discuss. To disperse or scatter.
Discutient. A medicine which scatters a swelling or tumor, or any coagulated fluid.
Diuretic. A medicine which increases the flow of urine.
Dolor. Pain.
Drastic. Powerful; efficacious.
Duodenum. The first of the small intestines.
Efflorescence. Eruption, or redness of the skin, as in measles, small pox, &c.
Effluvia. Exhalations, as the odor of plants, or the malaria from decayed animal or vegetable substances.
Electuary. A medicine composed of sugar or honey and some powder or other ingredient.
Eliminating. Discharging or throwing off.
Emaciation. Gradual wasting of the flesh; leanness.
Emesis. A vomiting.
Emetic. Any medicine which produces vomiting.
Emmenagogue. A medicine which promotes the menstrual discharges.
Emollient. A softening application which allays irritation.

Emulsion. A soft, milk-like remedy, as oil and water mixed with mucilage or sugar.
Enema. An injection.
Enteritis. An inflammation of the intestines.
Entozoa. Intestinal worms.
Epidemic. A prevalent disease.
Epidermis. The outer skin.
Epigastric. Pertaining to the upper and anterior portion of the abdomen.
Epileptic. Affected with epilepsy, or falling sickness.
Epispastic. An application for blistering.
Erosion. The act or operation of eating away.
Errhine. A medicine for snuffing up the nose to promote a discharge of mucus.
Eructation. The act of belching forth wind from the stomach through the mouth.
Eruption. A breaking out of humors on the skin.
Escharotic. Caustic; an application which sears or destroys the flesh.
Evacuant. A medicine which promotes the secretions and excretions.
Evacuate. To empty; to discharge from the bowels.
Exacerbation. An increase of violence of disease.
Exanthema. Such eruptive diseases as are accompanied by fever.
Excitant. A stimulant.
Excoriate. To gall; to wear off or remove the skin in any way.
Excrescence. A preternatural protuberance, as a wart.
Excretion. Useless matter thrown off from the system.
Exotic. Introduced from a foreign country.
Expectorant. A medicine which promotes a discharge of phlegm or matter from the lungs.
Expectoration. The act of discharging phlegm by coughing and spitting.
Expiration. The act of throwing out the air from the lungs, as in breathing.
Extravasation. Effusion; the act of forcing or letting a liquid out of its containing vessels.
Exudation. A sweating.
Feces (*Excrements*). The discharge from the bowels at stool.
Fauces. The back part of the mouth.

Febrifuge. Medicines that cure a fever, producing sweating.
Febrile. Indicating fever, or pertaining to fever.
Fetid. Having a strong and offensive smell.
Fetus. The child in the womb.
Fiber or *fibre.* A fine, slender substance which constitutes a part of the frame of animals; a thread.
Fibril. The branch of a fiber; a very slender thread.
Filment. A thread; a fiber
Filter. A strainer.
Filtration. Straining; the separation of a liquid from the undissolved particles.
Fistula. A deep, narrow, crooked ulcer.
Flaccid. Soft and weak; lax; limber.
Flatulency. Wind in the stomach and intestines, causing uneasiness, and often belching.
Flexible. Not stiff, yielding to pressure.
Flux. An unusual discharge from the bowels.
Fomentation. Bathing by means of flannels, dipped in hot water or medicated liquids.
Formula. A prescription.
Fundament. The seat; the terminating part of the large intestines.
Fungus A sponge-like excrescence, as proud flesh.
Gangrene. Mortification of living flesh.
Gargle. A wash for the mouth and throat.
Gastric. Belonging to the stomach.
Gland. A soft, fleshy organ for the secretion of fluids, or to modify fluids which pass through them.
Hectic. Habitual; an exasperating and remitting fever, with chills, heat, and sweat.
Hematosis. A morbid quantity of blood.
Hemoptysis. A spitting of blood.
Hemorrhage. A flux or discharge of blood from the nose, lungs, &c.
Hemorrhoids. The piles.
Hepatic. Pertaining to the liver.
Herbaceous. Pertaining to herbs.
Hereditary. That has descended from a parent.
Hernia. A rupture and protrusion of some part of the abdomen.
Herpes. An eruption of the skin; tetter, erysipelas, ringworms, &c.

Hydragogue. A purgative that causes a watery discharge from the bowels.
Hydrogen. A constituent of water, being one ninth part.
Hydrogen Gas. An aeriform fluid; the lightest body known. It is fatal to animal life.
Hydrophobia. A dread of water; the rabid qualities of a mad dog.
Hygiene. That of restoring or preserving the health without the use of medicine.
Hypochondriac. A person afflicted with debility, lowness of spirits, or melancholy—or in other words, with the blues.
Hysterical. Troubled with fits or nervous affections.
Idiopathy. A morbid condition not produced by any other disease.
Idiosyncrasy. Peculiarity of constitution or temperament; as being peculiarly susceptible of certain extraneous influences, and hence liable to certain diseases which others would escape from.
Ileum. The lower part of the small intestines.
Incrassation. Thickening.
Incubus. The nightmare.
Indigenous. Native.
Indurated. Hardened.
Infection. Communication of disease from one to another; contagion.
Inflammation. Redness and swelling of any part of the body, with heat, pain, and symptoms of fever.
Inflated. Filled or swelled with air.
Infusion. A medicine prepared by steeping either in cold or hot water.
Ingestion. Throwing into the stomach.
Injection. A liquid medicine thrown into the body by a syringe or pipe; a clyster.
Inoculation. Communicating a disease to a person in health by inserting contagion in his flesh.
Inspiration. Drawing or inhaling air into the lungs.
Inspissation. Rendering a fluid substance thicker by evaporation.
Integument. The skin, or a membrane that invests a particular part.
Intermittent. Ceasing at intervals.

Lanceolate. Oblong, and gradually tapering toward the outer extremity.
Larynx. The upper part of the windpipe.
Laxative. A gentle purge; a medicine that loosens the bowels.
Lesion. A hurt; a wound.
Lethargy. Unusual or excessive sleepiness.
Leucorrhea. The whites.
Levigated. Ground or made fine.
Ligature. A bandage.
Liniment. A stimulating ointment.
Lithontryptic. A solvent of stone in the bladder.
Lithotomy. The cutting for stone in the bladder.
Lubricate. To make smooth.
Lumbago. A pain in the loins or small of the back.
Lumbar. Pertaining to the loins.
Macerate. To dissolve in water.
Malaria. Bad air; air which tends to produce disease.
Manna. A laxative medicine obtained from the flowering ash.
Menstruum. A dissolvent. Any liquid used to extract the medical virtue from solid substances.
Morbid. Diseased; not sound or healthful.
Morbific. Causing disease.
Mucilage. A slimy substance.
Narcotic. A stupefying, sleep-producing medicine, often administered to relieve pain.
Nausea. Any sickness accompanied with an inclination to vomit.
Nervine. A medicine that operates on the nerves.
Ossify. To change flesh or other soft matter into hard bone.
Oval. Egg-shaped.
Oxygen. A constituent part of the atmosphere.
Palpitation. A violent beating of the heart.
Panacea. A universal medicine.
Paralysis. A loss of power of motion in any part of the system.
Paroxysm. A fit of any disease
Pathology. The science of the cause, symptoms, and nature of disease.
Pectoral. Pertaining to the chest; medicines for the cure of chest diseases.
Pericardium. A membrane inclosing the heart.
Plethora. Fullness or excess of blood.

Pleura. A thin membrane which lines the inside of the chest, and invests the lungs.
Polypus. A pear-shaped tumor.
Prolapsus. A falling down of some part of the body.
Pulmonary. Pertaining to, or affecting the lungs.
Pungent. Sharp; piercing; biting; stimulating.
Purgative. A medicine that evacuates the bowels.
Rectum. The last part of the large intestines.
Restorative. A medicine for restoring vigor and strength.
Rubefacient. An application which produces redness of the skin.
Saccharine. Having the qualities of sugar.
Saliva. Spit, or spittle.
Sanative. Healing, or tending to heal.
Scirrhous. Hard; knotty.
Scorbutic. Pertaining to scurvy.
Scrotum. The pouch containing the testicles.
Secretion. The act of producing from the blood substances different from the blood itself.
Sedative. A quieting, soothing medicine.
Serous. Thin, as the watery part of the blood, or of milk.
Sinew. That part which unites a muscle to a bone.
Slough. To separate from the sound flesh.
Solution. A liquid in which a solid substance has been dissolved.
Solvent. Having the power of dissolving solid substances.
Spasm. A violent but brief contraction of the muscles or fibers.
Spasmodic. Consisting in, or relating to, spasms.
Spleen. The melt.
Stimulant. An exciting agent.
Styptic. A medicine which coagulates the blood and stops bleeding.
Sudorific. A medicine which produces sweating.
Suppurate. To form purulent matter or pus, as a boil.
Tepid. Moderately warm.
Thorax. The chest.
Tincture. A medicine dissolved in alcohol or proof spirits.
Trachea. The wind-pipe, or breathing passage.
Tubercle. A pimple; a swelling, or tumor.
Tumefaction. The act of swelling, or forming a tumor.

Ulcer. A sore discharging pus.
Ureter. A duct, or tube, through which the urine passes from the kidneys to the bladder.
Urethra. The canal through which the urine passes from the bladder and discharges it.
Venery. Intercourse of sexes.
Vermifuge. A worm-destroying medicine.
Vertigo. Dizziness, or swimming of the head.
Virus. Contagion, or contagious matter; poison.
Viscid. Sticky, tenacious; like glue.
Vitiate. To injure; to impair; to spoil.
Volatile. Wasting away on exposure to the atmosphere.
Vulnerary. A medicine used for the cure of wounds.

INDEX.

THE HORSE AND HIS DISEASES.

Age of a Horse,	169
American Trotting Horse	13
Arabian Horse,	10
Apoplexy,	120
Blood-letting,	76
Bots,	111
Canadian Horse,	11
Cleaveland Bay,	12
Conestoga Horse,	12
Clydesdale Horse,	13
Catarrhal Inflammation of the Lungs,	96
Catarrh, or Cold,	96
Chronic Cough,	98
Curb,	141
Corns,	144
Cribbing,	162
Diseases of the Lungs,	94
Diseases of the Stomach,	104
Diseases of the Bowels,	106
Diabetes,	117
Diseases of the Skin,	128
Diseases of the Foot,	141
Ethan Allen,	28
Excretories the only Outlets,	50
Efforts of Nature to Remove Disease,	53
Feeding, Exercise, and Grooming,	155
Flora Temple,	37
Fevers in General,	83
Fistula in the Withers,	134
Founder,	145
Farcy,	126
General Management of the Horse	155
General Pathology,	47
George M. Patchen,	31
General indications of cure,	57
Grease,	129
Heaves,	101
Height of Trotting Horses,	41
Hidebound,	128
History of the Horse,	7
Inflammation of the Lungs,	94
" " Stomach,	104
" " Bowels,	106
" " Liver,	113

Inflammation of the Kidneys, ... 115
" " Bladder, ... 116
" " Brain, ... 118
" " Eye, ... 127
Immediate Cause of Disease, ... 84
Jaundice, ... 114
Morgan Horse, ... 10
Mode of Nature in Curing Disease, ... 58
Mercury, ... 58
Minerals in General, ... 74
Mode of Producing Perspiration, ... 90
Megrims, ... 119
Mange, ... 128
Norman Horse, ... 11
Nasal Gleet, ... 123
Performances of American Trotters, ... 20
Physician only the Handmaid of Nature, ... 58
Quitter, ... 144
Race Horse, ... 9
Remote Cause of Disease, ... 48
Ringbone, ... 139
Rules for choosing a Trotting Horse, ... 41
Rupture of the Suspensory Ligament, ... 139
Secret of Horse-Taming, ... 165
Simple or Primary Fever, ... 91
Spasmodic Colic, ... 108
Spavin, ... 141
Splints, ... 140
Strains, ... 136
Stomach Staggers, ... 105
Strain of the Flexor Tendon, or Back Sinew, ... 138
" " Hip or Whirl-bone Joint, ... 137
" " Shoulder, ... 136
" " Stifle, ... 137
Strangles, or Horse Distemper, ... 97
Surfeit, ... 128
Symptomatic Fever, ... 93
Tetanus, or Lock-jaw, ... 121
Thick Wind, ... 99
Thorough-pin, ... 140
Thrush, ... 145
Training the Trotting Horse, ... 42
Tumors on the Liver, ... 115
Unequal Circulation of the Blood, ... 49
Virginia Horse, ... 13
Water-Farcy, or Yellow Water, ... 126
Wild Prairie Horse, ... 13
Windgalls, ... 140
Wind-sucking, ... 164
Worms, ... 110
Wounds, ... 130

DISEASES OF HORNED CATTLE.

Abortion, or Losing the Calf,	201
Crook in Cattle,	191
Cough, or Hoose,	192
Clover Fever, or Bloating,	202
Choked with Apples or Potatoes,	202
Dropsy, or Water Tympany,	193
Diarrhea, or Looseness,	194
Flatulent Colic,	196
Falling of the Womb,	199
Garget, or Cake in the Bag,	200
How to Assist a Cow in Calving,	197
Horn Distemper,	201
Inflammation of the Kidneys,	185
" " Liver,	186
" " Lungs,	186
" " Bladder,	188
" " Stomach,	189
Inflammation of the Brain,	190
Losing the Cud,	204
Milk Fever,	181
Murrain, or Plague,	205
Quinsy,	189
Red Water, or Bloody Urine,	195
Simple Fever,	183
Symptomatic Fever,	184
Scrofula, or Scab,	203
Worms in the Tail,	204
Yellows, or Jaundice,	192

DISEASES OF SHEEP.

Abortion, or Losing the Lamb,	222
Apoplexy,	211
Catarrh, or Cold,	213
Diseases of Sheep,	209
Dropsy,	222
Diarrhea,	215
Dropsy, or Red Water,	216
Dysentery,	216
Erysipelas,	225
Falling of the Womb,	223
Foot-rot,	228
Fouls,	231
Garget,	223
Grub in the Head,	112
Hoove, or Distention of the Stomach,	213
Inflammation of the Lungs,	221
Maggots,	227
Pelt-rot,	226
Poison from Herbs,	217

Sore Mouth, ... 221
Stretches, ... 214
Sturdy, or Staggers, ... 211
The Scab, or Itch, ... 224
The Rot, ... 218

DISEASES OF SWINE.

Bite of Viper or Dog, ... 238
Diseases of Swine, ... 232
Inflammation of the Lungs, ... 237
Kernels, ... 236
Lethargy, ... 236
Loathing of Meat, ... 237
Measles, ... 235
Pestilence, or Plague, ... 237
Quinsy, ... 235
Shaking, ... 238
Staggers 240
Stoppage of the Issues 236

MEDICINES.

Alteratives, ... 148
Antiseptics, ... 152
Antispasmodics, ... 150
Balsams, ... 175
Clysters, ... 153
Cathartics, ... 170
Diuretics, ... 176
Drenches, ... 169
Discutients, ... 152
Explanations of Medical Terms, ... 265
Fomentations, ... 152
Injections, or Enemas, ... 178
Laxatives, ... 149
Liniments, ... 173
Poultices, ... 153
Powders, ... 167
Pills, ... 174
Poisons and Antidotes, ... 260
Recipes, ... 241-259
Rubefacients, ... 151
Rowels, ... 167
Sudorifics, ... 149
Styptics, ... 152
Sedatives, ... 152
Stimulants, ... 150
Tonics, ... 150
Tinctures, ... 177
Vermifuges, ... 176
Vesicants, or Blisters, ... 151

www.ingramcontent.com/pod-product-compliance
Lightning Source LLC
Chambersburg PA
CBHW032044230426
43672CB00009B/1458